For more information visit our web site:
www.oup.co.uk/general/vsi/

John A. Matthews and David T. Herbert

GEOGRAPHY

A Very Short Introduction

OXFORD
UNIVERSITY PRESS

OXFORD
UNIVERSITY PRESS

Great Clarendon Street, Oxford OX2 6DP

Oxford University Press is a department of the University of Oxford.
It furthers the University's objective of excellence in research, scholarship,
and education by publishing worldwide in

Oxford New York

Auckland Cape Town Dar es Salaam Hong Kong Karachi
Kuala Lumpur Madrid Melbourne Mexico City Nairobi
New Delhi Shanghai Taipei Toronto

With offices in

Argentina Austria Brazil Chile Czech Republic France Greece
Guatemala Hungary Italy Japan Poland Portugal Singapore
South Korea Switzerland Thailand Turkey Ukraine Vietnam

Oxford is a registered trade mark of Oxford University Press
in the UK and in certain other countries

Published in the United States
by Oxford University Press Inc., New York

British Library Cataloguing in Publication Data

Data available

Library of Congress Cataloging in Publication Data

Matthews, John A. (John Anthony), 1947–
Geography: a very short introduction / John A. Matthews and David T. Herbert.
p. cm.
Includes bibliographical references.
ISBN 978-0-19-921128-9
1. Geography. I. Herbert, David T. II. Title.
G70.M376 2008
910–dc22   2008000176

ISBN 978-0-19-921128-9

5 7 9 10 8 6 4

Typeset by SPI Publisher Services, Pondicherry, India
Printed in Great Britain by
Ashford Colour Press Ltd, Gosport, Hampshire

For John's mother-in-law, Annie D'Sa (M.F.M.I.L.) in Nairobi, Kenya; and David's grandchildren, Sion and Ella in Cardiff, *and* Isabel and Rosie in Bristol.

Contents

Preface

The aim of *Geography: A Very Short Introduction* is to provide
a succinct and lively, yet authoritative, account of the nature
of geography as a field of study. For most people, the term
'geography' has an instant, if over-simplified, meaning. Different
countries in the world, rivers, mountains, and capital cities, and
their location on maps, are often among the first things that
come to mind. If a contestant in a popular quiz show chooses the
category of geography for his or her questions, these are often
the questions posed. Geography is of course far more complex
than this inventory of factual material. Its subject matter is
extremely varied, its concepts are many and well developed and
its methodologies are rigorous. It is a broad church with a range
of interests and involvements that is often surprising. Modern
geography has come a long way from simple descriptions of places
and landforms, and it is this modern face that we need to portray.
One stance we wish to emphasize is the centrality of geography to
many of the big issues that beset the modern world. These range
from global warming and other aspects of environmental change
to the spatial incidence and spread of diseases such as HIV and
AIDS. Geographers have the skills and experience to be involved
in teams that address issues of this kind.

Geography has always fallen into two parts, physical and human.
Physical geographers study the Earth's surface as a physical entity

with its landforms, vegetation cover, soils, climatic variation, and so on. Human geographers are concerned with the ways in which people occupy the Earth's surface, their movements and settlements, and their perceptions and use of the land, resources, and space. Out of this duality rises one of the strengths of geography: its ability to act as a bridge between nature and society. We will outline the original 'Geographical Experiment' that rested on this synergy between nature and culture, and recognized the unique position of geography between the sciences and the humanities. The integration of geography as a single discipline, which was explicit and strong through the earlier decades of the 20th century, has became less so in modern times. Greater specialization means that physical and human geographers tend to follow different agendas and refer to different sets of academic literature and scholarship. These trends will be closely discussed, and one of our stances will be to advocate the continuing value and strength of integrated geography.

We have acknowledged key sources but owe particular thanks to Sietse Los for preparing the satellite images of the Aral Sea; to Giles Young, who climbed the mountain to take the photograph of Storbreen glacier foreland; and to Nicola Jones and Anna Ratcliffe for their sterling work on drawing, modifying, or otherwise preparing all the illustrations in their final form. We are also grateful to Andrea Keegan for her insights and suggestions.

List of illustrations

Sources are listed in full in the References section at the end of the book.

The publisher and the author apologize for any errors or omissions in the above list. If contacted they will be pleased to rectify these at the earliest opportunity.

Chapter 1
Geography: the world is our stage

What is geography? An initial answer to this question can be obtained by discussing the essence of geography and identifying its distinctive characteristics. Some of this essence was captured by T. S. Eliot when he wrote:

> We shall not cease from exploration
> And the end of all our exploring
> Will be to arrive where we started
> And know this place for the first time.

T. S. Eliot, Little Gidding (1942)

One essential characteristic of geography that emerges in this quotation is the desire to discover more about the world in which we live; to record its many parts, ceaselessly to encounter the strange and new, and yet always to return to our roots, to the place we have chosen to call home. So much of the history of geography and so many of the great landmarks in the history of civilizations have had their beginnings in this thrust to explore and to understand.

Echoes from the past

A brief look at the historical record shows that geography has always been important. Writing in 58 BC, Julius Caesar recorded some of the characteristics of northern Europe that were relevant to a Roman general:

> The whole of Gaul is divided into three parts, of these one is inhabited by the Belgae, a second by the Aquitani, and a third by a people called Celts in their own language and Gauls in ours. Each differs from the others in language, customs and laws …
>
> Geography confines the Helvetii in all directions. On one side the broad and deep Rhine separates the Helvetian country from the Germans; on another the lofty Jura range lies between the Helvetii and the Sequani; and on a third side Lake Geneva and the Rhone separate the Helvetii from our province.
>
> Julius Caesar, *Gallic War* (58 BC)

His description is demarcating territories, identifying key boundary markers and recognizing their human distinctiveness: what would now be termed regional geography is being produced. Equivalent regional descriptions today might identify differences within and between the European Union, America, China, the former Soviet bloc, or the Islamic world.

The earliest geographies were often descriptions of lesser-known parts of the world to inform a 'home population'. Herodotus, for example, wrote about the different places in the Roman world, of their natural environments and their cultural occupation. The early mapmakers charted shores, rivers, and mountains and presented navigational and informed depictions of the Earth's surface. During the heady days of exploration and discovery, the geographies were essential means of communication between the explorers and the general public, including their sponsors. The Royal Geographical Society, founded in 1830 in London, became a key forum for the reporting and dissemination of the great expeditions of that time. These reports captured the popular imagination, and explorers such as Livingstone, Stanley, Burton and Speke, and Nansen, Shackleton, Scott (Figure 1), and

1. **Robert Falcon Scott's party at the South Pole on 18 January 1912: (left to right) Dr Wilson, Captain Scott, Seaman (Taff) Evans, Dr Oates, and Lieutenant Bowers. The negative of this photograph was found inside the tent in which Scott and three of his companions perished**

Amundsen became icons of the age of discovery along with sailors, such as Columbus, Vasco da Gama, and Cook, who reached out to distant parts of the world.

The following is part of the account given by David Livingstone when, in 1856, he sailed down the Zambesi River and named Victoria Falls:

> After twenty minutes' sail from Kalai, we came in sight, for the first time, of the columns of vapour, appropriately called 'smoke', rising at a distance of five or six miles, exactly as when large tracks of grass are burned in Africa. ... The whole scene was extremely beautiful; the banks and islands dotted over the river are adorned with sylvan vegetation. ... There, towering over all, stands the great burly baobab, each of whose enormous arms would form the trunk

of a large tree....The falls are bounded on three sides by ridges 300
or 400 feet in height which are covered with forest, with red soils
appearing among the trees.

David Livingstone,
Missionary Travels and Researches (1857)

Livingstone's main purpose in Africa was his missionary activities,
but others at that time were driven by commercial, political, or
scientific considerations. Lands previously unknown to Europeans
were being discovered and new facts about the physical make-up
of the Earth's surface, its landscapes, environments, and
resources, were being made known. The process of geographical
exploration, its reporting and depiction of the Earth's surface, had
major significance for the history of science. Indeed, for most of
historical time, geographical progress has been indistinguishable
from that of science in general.

The first lines of Charles Darwin's account of his scientific
research during the voyage of the *Beagle* were:

> After twice being driven back by heavy south-western gales, Her
> Majesty's ship Beagle, a ten-gun brig, under the command of
> Captain Fitz Roy, R.N., sailed from Devonport on 27th December
> 1831. The object of the expedition was to complete the survey of
> Patagonia and Tierra del Fuego, commenced under Captain King
> in 1816 to 1830 – to survey the shores of Chile, Peru and some
> islands in the Pacific – and to carry out a chain of chronometrical
> measurements round the World.

Charles Darwin, *The Voyage of the Beagle* (1845)

This was the journey that inspired Darwin to produce his theory
of evolution that changed the scientific world. That theory
was inspired by the geographical variations in species that he
observed, especially those encountered in the Galapagos Islands,
but the primary purpose of the voyage was to produce maps and
charts and descriptions of this part of the world. Those maps had

a purpose. Mostly their purpose was purely functional – to aid navigation, produce an accurate record, and pave the way for later expeditions – and this was the practical face of geography at that time.

This kind of fundamental geography was a practical science, characterized by its empiricism and fact-finding that helped build up our knowledge of the world. Much of the exploration and accompanying mapmaking was politically motivated. It was strongly linked to imperialism and colonies and to attempts to extend the power of particular states and organizations. Maps were a means of depicting claims to territory and demarcating growing spheres of influence. Reminiscing on his Victorian childhood, the author Stuart Cloete wrote in his autobiography:

> From pole to pole the Union Jack whipped in Arctic gales or sagged in the tropical heat of this empire upon which the sun never set.... Britannia rules the waves. Pax Brittanica was a reality. London was the centre of the world: in children's school atlases country after country, whole continents, were painted red with unquestioned British dominion.

Stuart Cloete, A Victorian Son (1923)

As with the Roman Empire before it, there was subsequent decline and fall, but the scramble for lands and colonies was aided by and contributed to the science of geography. The last quarter of the 19th century was the Age of Imperialism. In 1875, 10% of Africa was in foreign hands; by 1900, this figure had risen to 90%. Britain led the way in redrawing the map of Africa, but France, Germany, Belgium, Portugal, Spain, and Italy all had a part to play. Every representation of the world in the form of a map and every journey into new lands reflected deeper agendas. The questions were of extending influence, establishing control, and presenting an image that suited a particular purpose. Geography, then, has always been concerned with the 'where' of things and their relationships in space. Mapmakers sought to depict this quality, driven by the rigours of their scientific method, but the users and perhaps commissioners of maps were alive to their power to shape the Earth's surface.

Another moment in time that reveals the essence of geography is provided by the Mormons' 1,300-mile pioneer trail across the United States led by Brigham Young in 1847. After months of hardship, the lead group came through a gap in the Wasatch Mountains and looked down upon Great Salt Lake. A monument marks the point where Brigham Young reached this view, looked down, and said, 'This is the place'. It became the place for the site of Salt Lake City, the title of Utah's state song, and the founding of the Mormon cultural region; a place endowed with meaning and symbolism. The significance of such a place, which varies through time depending on the people who see, interpret, and use it, is as much a part of geography as the factual descriptions of the Earth's surface that cartographers seek to portray.

A further important dimension of the essence of geography was highlighted by the geographer George Perkins Marsh:

> There are parts of Asia Minor, of North Africa, of Greece, and even of Alpine Europe, where the operation of causes set in action by man has brought the face of the Earth to a desolation almost as complete as that of the Moon. ... The Earth is fast becoming an unfit home for its noblest inhabitant.
>
> George Perkins Marsh, *Man and Nature* (1864)

His quotation points to an abiding concern of geography with the natural environments of the Earth's surface and with the modifications brought about by human actions. Both positive and negative human impacts have always been a feature of the exploitation of resources, particularly through the use of fire and other types of technology. It is important to realize that the negative impacts are not merely the inadvert ones associated with recent global warming; they have often been produced by reckless and unthinking actions.

Geography is everywhere

Today, geography impinges upon our everyday lives in a multitude of ways; the expression 'geography is everywhere' is intended to reflect that special quality. Everything has a location on the surface of the Earth whether it is expressed in terms of latitude and longitude, some form of spatial grid-referencing system, or merely as distance from home or school or work. We move on the Earth's surface from one geographical location to another. Some of our journeys are short and frequent, such as the daily trip to work or school, others are longer and infrequent, such as vacation travel or visits to relatives who live some distance away.

Again, when we visit a supermarket or shopping centre, we find goods and commodities that have been brought from many different environments and parts of the world. There are bananas from the Caribbean, citrus fruits from Florida and South Africa, and a range of wines from France, Spain, Chile, California, Australia, and New Zealand. All of these provide links with different parts of the world and their geographies. There are other dimensions to these linkages. The large supermarkets, for example, set standards for safety, quality, and ethical aspects of food; it has been argued that supermarkets are now so powerful that 'food governance' has echoes of imperial governance. As we buy, use, and dispose of commodities, goods, or services, these actions connect us to other people and other places in ways that may be beyond our imaginings. If you walk through one of the major cities such as London, Paris, or New York, there are people to meet who have travelled from many different parts of the world; some are tourists or visitors on short stays, others are immigrants or refugees seeking a new life. Finally, we live our lives in well-marked spaces such as home, neighbourhood, city or town, region and country. These are all known geographical places; the territories that assume great importance in our lives. Geography, then, is everywhere and the study of geography examines these locations, connections, territories, environments, and places, and seeks to understand their significance.

The subject matter of geography is the Earth's surface, including the envelope of atmosphere immediately above it, the structures that lie immediately below it, and the social and cultural environments contributed by the people who occupy it. Common definitions of geography capture many of these qualities, albeit in very abrupt forms. Thus, geography as the 'where' of things is one catchphrase. 'Geography is about maps and history is about chaps', is another. Again, geography tells us about the world and its places. Most would agree that in an increasingly interdependent and connected world beset with problems of global significance, an understanding of its geography is essential. Current big issues such as global warming, environmental change, natural hazards, flows of refugees, rising levels of pollution, the rapid onset of epidemics, and burgeoning conflicts all have considerable geographical dimensions.

Emergence of geography as a university discipline

Whenever a new discipline establishes itself in universities, there are always problems of identity to resolve, and the story of geography is unexceptional in this respect. Mapping in geographical space goes back a long way in time so one basic principle of geography belongs to the distant past. Similarly, essential geographical concepts can be found in the writings of Greek philosophers, Roman historians, and Sumerian cartographers. Geography, with its empirical matter-of-factness, was a discernible element of the growth of knowledge, but its various concepts were not drawn together into an integrated subject area.

It is not until the 16th century, for example, that there is evidence of the coherent practice of geography in British universities. This practice was contained in a variety of schemes of study and reflected a wealth of intellectual traditions and established disciplines. Relevant learned societies, especially the Royal Geographical Society (RGS), offered support but tended to be strongly focused on the historic priorities of mapmaking, discovery, and exploration. Major changes were in the air in the 19th century. In the 'Age of Empires', maps acquired new meanings and the voyages and journeys of discovery had scientific as well as political interest. Many had seen Darwin's work on evolution by natural selection as the catalyst for studies of the geography of the natural environment. More directly, Halford Mackinder, the first Professor of Geography at Oxford, developed his 'Geographical Experiment', which involved the integration of the study of society and the environment, and the maintenance of culture and nature, under one umbrella. This defined geography at that time and set the challenge of understanding the relationships between these two principal components of the Earth's surface.

Developments in the United Kingdom did not occur in isolation. Alexander von Humboldt and Karl Ritter led movements towards a new geography in Germany, with the former emphasizing the Earth-surface features that created natural landscapes and the latter arguing for the recognition of regions in the world as the home of man. European ideas on the impact of environment on people sparked the debate on determinism that extended to American geographers. French geographers were strongly interested in cultural landscapes and regions that reflected traditions and ways of life. All of these new ways of thinking about people, environments, and the meanings of landscapes evolved during the 19th and early 20th centuries. They were part of the intellectual ferment that followed the new sciences and the lateral thinking of the late Enlightenment.

The Geographical Experiment gave geography the chance to establish itself as a university discipline. The breadth of its terms of reference was both a strength and a weakness. The strength was that it included nature and culture and their relationship, a concept no other discipline had claimed. This breadth remains a contested topic in modern geography despite the opportunities it presents of ever-increasing relevance. The weakness is the spread of interest over such a wide field and an 'anything goes' mentality. This weakness becomes most apparent when different parts of the discipline relate to different intellectual traditions. The touching points then become very few or non-existent. It is fair to say that most physical geography today is evolving within the research framework of the natural and mathematical sciences, whereas most human geography draws upon and interrelates with the traditions of the humanities and social studies. It is possible to recognize a definite lacuna in which physical and human geography interact, but for many this is a minority interest.

Geography is now a well-established university discipline. It is a common presence in European universities and is also widely found as both undergraduate and postgraduate programmes in most parts of the world. The International Geographical Union lists members in 75 different countries including, for example, 27 in Japan, 14 in South Africa, 10 in China, 5 in India, 4 in Peru, and 1 each in Morocco, the Philippines, Sudan, and Tanzania. Most departments are found in developed countries. The 2001 research assessment exercise for institutions of higher education in the United Kingdom recorded entries from 60 institutions comprising over 450 research-active academics. Current listings of universities and colleges offering degree schemes in geography in the United States show 217 institutions, and there are a further 42 in Canada. The current penchant is for mergers of geography departments with others into schools with titles such as Earth, human, or environmental sciences/studies. Many of these changes are fairly cosmetic, as geographical research and undergraduate studies continue in these new units, and students wishing to study geography will easily find degree schemes that fit their aspirations.

Geography's core concepts

How important, then, is geography? Hopefully, its fundamental importance is now clear and the next question is 'Where are its core concepts?'

Geography has always been involved in the analysis of *space* and this provides the first core concept. Geographical space comprises location, or where we are on the Earth's surface in relation to geographical coordinates; distances measured in a variety of ways; and directions that complete the interrelationships of different locations on the Earth's surface. A key corollary of the focus on geographical space has been the ways in which the Earth's surface is depicted. Maps, cartography, and, most recently, satellite images, qualified by scale and forms of representation, are the working tools for much geographical analysis.

For the cartographer or surveyor, space is an absolute and the science is that of depicting it with the correct detail. Geographers have struggled with the basic problem of depicting a spherical Earth on a flat piece of paper and the development of map projections summarizes that process. That particular outcome is one of compromises; one opts for true distance or true area but both together are not attainable. The seminal map projection of Mercator in the 16th century, in which the points of the compass always maintained true direction, set the benchmark. Human geographers have found that space is often more usefully represented in relative terms. For someone wishing to visit a shop, for example, 20 miles is a major obstacle to walk but of much less significance if he or she has access to a car. Distances are mediated by accessibility and that can be contributed to by type of terrain

11

as well as the type of person; flat areas are easier to manage than steep slopes. It can be argued that linear measured distance *per se* is not meaningful unless it is qualified by conditions of this kind.

Place is another core concept in geography. Place is not independent of space because it involves an area or territory; it is a form of bounded space. Place can be applied at a variety of scales from a state or country to a neighbourhood or home area. Place therefore includes the search for boundaries, edges, and limits that contain a definable and recognized territory. When describing the differences between places, the focus may be on natural boundaries such as rivers or mountain ranges, but boundaries are also set by human decision-makers who may be intent upon identifying political states or arbitrating among disputed territories. The physical boundaries are not always unambiguous, and the lesson of history is that major disputes and conflicts can arise over the designation of relatively small parcels of land. Geography also includes the mental maps and images that define places subjectively. Residents of a neighbourhood, for example, might be asked to sketch the boundaries of their home area or to construct their mental map of the city in which they live. People attach special and often individual meanings to places, such as the place where they spent their childhood or the place they associate with some special event. Different people from different cultures may perceive and interpret the same area of the Earth's surface in different ways.

There are, in short, meanings attached to places: these meanings may be affective and emotional, which does not lend them easily to measurement. There is, for example, a growing interest in literary places that either provide the settings for fictional novels or were the locations where writers lived and worked. Such places now attract many visitors who are at least as interested in the fictional settings and the characters that inhabited them as they are with the real lives of the authors. Haworth in Yorkshire, for example, where the Brontë family lived, attracts many visitors, but much of the attraction of the nearby Moors resides in the fact that the fictional characters of Heathcliff and Catherine Earnshaw walked there. Similarly, John Fowles imbued the Cobb at Lyme Regis with new meaning after the filming of his book *The French Lieutenant's Woman*. These examples illustrate the diversity of the concept of place: it can be a precise measured area on the ground, such as a field or a forest, but it can also be a subjective image or a well-defined location imbued with special meaning. A football field or other sporting venue can be regarded in both ways: it is measurable and precisely defined, but it can also be an iconic place, remembered as the scene of notable achievement and constituting part of the cultural life of thousands of people.

Environment is the third core concept for geography. In its most unambiguous interpretation, it is the natural environment, but that environment is occupied by people and in that sense it has a wider meaning. The environment, like place, encompasses human perceptions and aspirations as well as the biophysical characteristics that can be measured and monitored. The shape of the Earth's surface and the processes enacted upon it, both physical and human, are part of the essence of geography. Similarly, the reciprocal relationship between the natural environment and people always has been and remains a key question. The emphases have changed over time, from early ideas that suggested definitive environmentally determined limitations on people, to greater awareness of the human impact on the natural environment. Current issues of sustainability, custodianship of the environment, protocols to reduce holes in the ozone layer, and world summits to limit the use of carbon fuels all belong to the imperative to understand and manage this key relationship. Geographers would argue that they alone focus on the holistic view and that this is a view of increasing importance in a world where issues such as environmental change and globalization are becoming pressing.

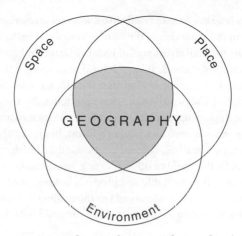

2. Three core concepts of geography: space, place, and environment. The essence of geography (shaded) is an integration of spatial variation over the Earth's surface with the distinctiveness of places and interactions between people and their environments

The three core concepts of space, place, and environment can be identified as the central concerns of geography (Figure 2); the bonds that hold the subject together and give it meaning.

Defining contemporary geography

Some shorthand working definitions of geography have already been mentioned, and it is useful now to consider some more formal definitions and to examine the degree of consensus that exists. Our own definition from an earlier publication is as follows:

> Geography is the study of the surface of the Earth. It involves
> the phenomena and processes of the Earth's natural and human
> environments and landscapes at local to global scales. Its basic
> division is between physical geography, which is unambiguously
> a science and analyses the physical make-up of the Earth's

surface.... and human geography, where the focus is on the human occupancy of this area.

D. T. Herbert and J. A. Matthews, The Encyclopaedic Dictionary of Environmental Change (2001)

This is a fairly lengthy quote but other more succinct definitions follow similar lines. The American geographer Edward Ackerman focuses on the idea of a system and interaction between people and nature:

> The goal of Geography is nothing less than an understanding of the vast interacting system comprising all humanity and its natural environment on the surface of the Earth.

E. A. Ackerman, 'Where is a Research Frontier?' (1963)

Another American geographer expresses the grand sweep of geography and its all-embracing character: here is a way to look at the Earth in all its diversity:

> Geography is the science of place, its vision is grand, its view is panoramic. It sweeps the surface of the Earth, charting its physical, organic and cultural domains.

Science, review of Harm de Blij's Geography Book (1995)

The last two definitions, again from American geographers, stress the scientific nature of geography and the interactive processes that operate in space and the environment.

> Geography is a discipline concerned with understanding the spatial dimensions of environmental and social processes.

G. F. White, Encyclopedia of Global Environmental Change (2002)

> Geography is the study and science of environmental and societal dynamics and society-environment interactions.

G. L. Gaile and C. J. Willmott, Geography in America (2003)

None of these definitions are too far apart. They invoke the core concepts and they stress the integrative role that gives geography its special meaning. At various times in the history of geography, a particular key concept may have been emphasized more than the others, but all three have co-existed and form the core of the subject. Similarly, within physical and human geography, more or less emphasis may be given to particular concepts for particular purposes, and their precise interpretation may vary. In modern times, there is greater awareness that the 'facts' of geography are not unambiguous; they mean different things to different people at varying points in time. The concept of place, for example, has moved from the simple demarcation of areas to a study of the affective values with which they are imbued. This acceptance of the ambiguities in the meaning of geography is in itself a positive attribute that opens up new lines of understanding.

Geography (Figure 2) should therefore be thought of as the nexus where the three core concepts – space, place, and environment – overlap. Space, place, and environment, as we have defined them above and as will be elaborated further in later chapters, are a necessary part of the discipline of geography. None is sufficient, on its own, to define geography. Hence the essence of geography is represented by the shaded area in Figure 2. Is there a single term to describe this integrated area? Possibly not, but the concept of landscape comes close to defining this nexus that is geography. There are two metaphors that help to illuminate this claim. The first is the idea that landscape has the qualities of a palimpsest. Literally, a palimpsest was a form of parchment that, before the advent of paper, was written upon many times. Although the aim was to erase the previous writing, it inevitably left its traces. A landscape can be seen in the same way. It has been written over many times by both physical and human processes, but traces of the past are still discernible. The second metaphor is provided by the French human geographer Vidal de la Blache, who likened landscape to a 'medal cast in the image of its people'. For him, the record of people's occupance of the land over long periods of time could be read from the study of landscape. The evidence might be, for example, an archaeological site, a pattern of fields, or a form of settlement. Landscape then approximates the nexus of geography. The study of regions as integrated parts of the Earth's surface that combine nature and culture can be viewed in the same way.

No other discipline focuses on the space-place-environment nexus. This has been the focus of geography throughout its history and still defines its role today. At the same time geography has developed. Much has changed in the ways particular concepts are interpreted and research is carried out. Figure 3 traces the broad path of these changes, the major phases through which geography has passed, and the divergences and tensions that have recently emerged. Phase 1 was the long period of time up to the mid-19th century when the explorers and mapmakers sketched out the properties of the known world. The beginning of the 20th century, phase 2, witnessed the establishment of an identity for the discipline of geography within universities founded on its bridging role between nature and culture. During the first half of the 20th century, phase 3 shifted the focus towards regional studies and human landscapes; and phase 4, dating from the middle decades of the 20th century, saw the clear emergence of subdisciplines within the broad categories of physical and human. Phase 5, which began during the late decades of the 20th century, brings us through to modern times and the increasing diversity of a wide field of study; the contemporary geography that we will seek to illuminate.

Within the practice of contemporary geography, many traditional components such as maps are still important, though satellite remote sensing, sometimes known as Earth Observation (EO),

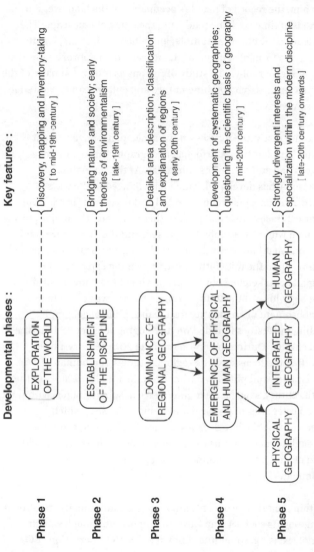

Developmental phases :

Phase 1 — EXPLORATION OF THE WORLD

Phase 2 — ESTABLISHMENT OF THE DISCIPLINE

Phase 3 — DOMINANCE OF REGIONAL GEOGRAPHY

Phase 4 — EMERGENCE OF PHYSICAL AND HUMAN GEOGRAPHY

Phase 5 — PHYSICAL GEOGRAPHY / INTEGRATED GEOGRAPHY / HUMAN GEOGRAPHY

Key features :

Discovery, mapping and inventory-taking
[to mid-19th century]

Bridging nature and society; early theories of environmentalism
[late-19th century]

Detailed area description, classification and explanation of regions
[early 20th century]

Development of systematic geographies; questioning the scientific basis of geography
[mid-20th century]

Strongly divergent interests and specialization within the modern discipline
[late-20th century onwards]

3. Five main phases in the development of geography and some of their key features

Geographical Information Systems (GIS), and other powerful quantitative methods have been added to traditional fieldwork and the comparative method. Established core concepts of space and place have been transformed, in human geography at least, by modern social and cultural theory. The need to understand the biophysical and human environments of people and their interactions is becoming increasingly urgent as issues of sustainability and the protection and preservation of planet Earth become imperative. As integration within geography as a whole has weakened, both physical and human geography have become more specialized and have adopted different approaches to many of their research problems. Most importantly, physical geography is asserting its scientific credentials, while human geography emphasizes critical theory, values, and ethics.

Modern geographical exploration and discovery are therefore different from the days of Christopher Columbus, David Livingstone, or Robert Falcon Scott, but are just as important. There are still expeditions, such as those run by the Royal Geographical Society to the Mato Grosso of Brazil in the 1960s, the Mulu rainforest of Sarawak in the 1970s, and the Wahiba Sands of the Sultinate of Oman in the 1980s. They are now commonly termed 'research projects', though their exploratory purpose remains. The aim of the Wahiba Sands Project, for example, was to examine the sand sea of the Wahiba Sands as a complete geo-ecosystem, including the sands themselves, the biological resources, and the people. Perhaps the main difference from the traditional model was that it led to a management plan for sustainable development.

Modern geography forms an essential component, not only of the natural and social sciences, but also of the humanities. There are still expeditions into the unknown, but geography has changed as what is 'known' has changed. Computers, laboratories, and libraries are now just as indispensable to geographers as maps and fieldwork. And for many behavioural and cultural geographers in particular, the *terrae incognitae* of the early explorers have been replaced by those of the human mind. However, Neil Armstrong's first footsteps on the moon show that there is still potential for traditional exploration; and perhaps, in this sense, space may indeed prove to be the 'final frontier'.

Chapter 2
The physical dimension: our natural environments

Study of the natural environment has always been an essential part of geography and forms the focus of this chapter. Physical geography may be defined as the natural environmental science of the Earth's surface. But what are its characteristics, and how did they develop? How does physical geography interact with the other sciences that investigate the Earth's natural environments, and what exactly is its special role?

Geo-ecosphere: the playing field

Thinking of the Earth's surface as the 'geo-ecosphere' – the narrow surface zone comprising all the landscapes of the Earth – is helpful in defining the overall scope of physical geography as depicted in Figure 4(A). The geo-ecosphere can be subdivided into six component spheres, each of which has attracted its own specialist physical geographers. Thus, the topography of the Earth's land surface (toposphere) can be seen as the focus for geomorphology; the totality of life on Earth (biosphere) is the focus for biogeography; and the lower layers of the gaseous envelope (atmosphere) are the focus for climatology. Other important spheres identified in Figure 4(A) are the pedosphere, which involves the soil cover of the Earth; the hydrosphere, which incorporates the liquid water in rivers, lakes, oceans, and

A

Atmosphere

GEO-ECOSPHERE

Topo-sphere

Cryosphere

Hydrosphere

Biosphere

Pedosphere

Lithosphere

B

GEO-ECOSPHERE   ANTHROPOSPHERE   NOÖSPHERE

4. The geo-ecosphere (shaded) – the subject matter of physical geography – including: (A) the natural geo-ecosphere and its component spheres; and (B) the human-influenced geo-ecosphere, or anthroposphere

groundwater; and the cryosphere, the world of snow, ice, and 'frozen' ground.

Other scientists are, of course, interested in these 'spheres'. The toposphere, for example, is influenced in a major way by the underlying rocks of the lithosphere, which is studied primarily by geologists, geophysicists, and geochemists. Atmospheric scientists – meteorologists, atmospheric physicists, and atmospheric chemists – study the atmosphere, including its upper layers, which are of minor interest to physical geographers. Botanists, zoologists, and ecologists investigate the biosphere; and pedologists, hydrologists, glaciologists, and others specialize in the other spheres mentioned above. What differentiates physical geography from these other scientific fields is the focus on spatial patterns in the landscape and their underlying dynamics from local, through regional, up to global scales. Local-scale patterns include, for example, hillslope form and the shapes of valleys, meandering rivers, woodland distribution, and urban climate. At regional scales, mountain ranges, major river basins, and climatic zones come to prominence; whereas at the global scale, global warming, deforestation and the loss of biodiversity, and interactions within the Earth-ocean-atmosphere system, are amongst the topics investigated.

Physical geographers investigate not only variation from place to place in the various spheres but also the interactions between the different spheres and their changes through time. El Niño events provide a good example of interactions at several scales in space and time. Named after the warm El Niño ocean current that appears off the coast of Ecuador and northern Peru just after Christmas, these periodic events begin in the tropics with intense warming of the surface waters of the equatorial Pacific Ocean and are propagated around the globe producing worldwide effects. Typical effects of an El Niño event in the northern hemisphere winter are mapped in Figure 5. They include droughts in Indonesia, eastern Australia, and southern Africa, while severe storms and flooding occur along the coast of Ecuador and Peru, and throughout the Gulf States of North America.

Although physical geography is defined by its emphasis on spatial patterns and spatial processes in the geo-ecosphere, human activity also plays a major role. The thin, surface 'skin' of planet Earth is the natural environment on which the human species is partially, if not entirely, dependent. What distinguishes humans

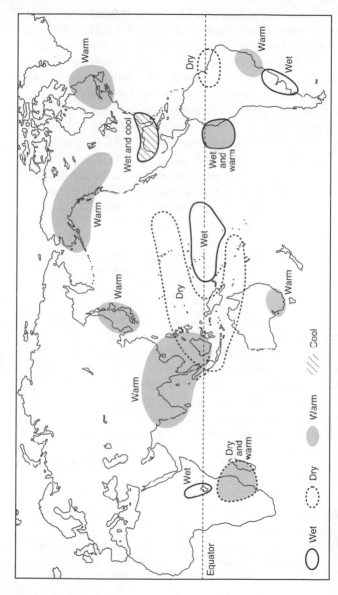

5. **Typical climatic anomalies during an El Niño event in the northern hemisphere winter**

from the rest of the biosphere, however, is the conscious ability to create their own cultural and technological environments. It is therefore possible to recognize a sphere of human mental activity, which has been termed the 'noösphere'. This is shown in Figure 4(B) alongside the natural geo-ecosphere. The overlap between these two spheres represents the human-modified geo-ecosphere, or anthroposphere. Herein lies an essential connection between physical and human geography, and it is in this sense at least that the natural environment can be regarded as the physical basis of geography.

As human impacts on the natural environment increase inexorably, it is more difficult to differentiate a natural geo-ecosphere from the anthroposphere. Most of the Earth's surface and its component spheres are impacted by both natural and human disturbances of various types. Agriculture now regularly affects around 45% of the Earth's terrestrial surface, forestry some 10%, transportation 5%, urban development 3%, and mineral extraction 1%. Even military activities affect, or have recently affected, an appreciable area, ranging from 1% of the USA to 40% of Vietnam. This does not mean either that *all* of physical geography is concerned with human environmental impacts, or that there is a physical environmental basis to *all* of human geography, but that the nature of the interaction must always be considered.

The early development of physical geography

Of the early founders, the most eminent proponent of physical geography as a scientific entity was undoubtedly the German polymath Alexander von Humboldt. On his many travels, he combined observations with measurements of temperature, pressure, and the Earth's magnetic field, and made generalizations about the geographical distribution of vegetation, global-scale patterns of temperature (depicted by isotherms on maps), the ways in which temperature falls and vegetation varies with increasing altitude (on Tenerife in the Canary Islands, for example), the alignment of volcanoes, and the course of ocean currents. In his major works, written around the middle of the 19th century, such as *Cosmos: A Sketch of a Physical Description of the Universe*, published in 1849, he emphasized not only relationships within the natural geo-ecosphere but also linkages to human societies. A year earlier, Mary Somerville, based at the University of Oxford, published *Physical Geography* and defined the subject as 'a description of the Earth, the sea and the air, with their inhabitants animal and vegetable, of the distribution of these organized beings and the causes of that distribution'.

Another major early influence was the publication of Charles Darwin's *The Origin of Species by Means of Natural Selection* in 1859. This work had profound effects on all the natural environmental sciences, including physical geography. Views that had regarded the geo-ecosphere as harmoniously integrated but essentially static had to accommodate an ever-changing, continually adjusting and evolving Earth's surface. Thus, Thomas Huxley's *Physiography*, published in 1877, while developing his theme with particular reference to the Thames Basin in south-eastern England, emphasized chains of causal connections between the various natural components of the landscape within an evolutionary framework.

By the early 20th century, the concept of a 'cycle of erosion', also termed the 'geographical cycle', was developed in particular by the American geographer William Morris Davis (see box). He used the idea that landforms represent various stages in a sequence from 'youth' to 'maturity' and 'old age'. This proved to be the dominant theory in geomorphology for the next half century. Although the geomorphological landscape was seen as the product of structure (the underlying geology), process (primarily running water eroding the surface), and stage (the age of the landscape and hence its stage within the cycle), it was the emphasis on

The cycle of erosion

William Morris Davis envisaged a cycle of erosion that was initiated by uplift of the land surface. This was followed initially by rapid incision by rivers and later by valley-floor widening. The overall pattern was of curved (convexo-concave) slopes declining in angle and eventually terminating in a landscape of low relief known as a peneplain (until land uplift rejuvenated the landscape and the cycle began again). The essentials of his model are shown in Figure 6(A). Effects that could not be fitted into the so-called 'normal cycle', such as the landforms produced by glaciation, were viewed as 'climatic accidents', though different cycles of erosion were later proposed for regions with climates that differed from the largely temperate and fluvially dominated regions of the USA and Europe where the original model was developed. One of these alternative models, which was thought more appropriate for the semi-arid regions of southern Africa, advocated the parallel retreat of slopes rather than Davisian slope decline (Figure 6(B)). According to this model, steep slopes and extensive remnants of the initial land surface remained late into the cycle.

While they did focus the thoughts of geomorphologists on the idea of systematic change over time, such 'cycles' were also a constraint. These theoretical models were too simple in relation the complex evolution of real landscapes. In particular, landscapes do not stay stable long enough for the completion of the full cycle because of both the tectonic forces controlling uplift and the environmental changes affecting Earth-surface processes. Modern ideas on landscape evolution give much greater attention to how landscapes react to changing conditions, to rates of landscape change in the past, and to the response of the landscape to likely future environmental change.

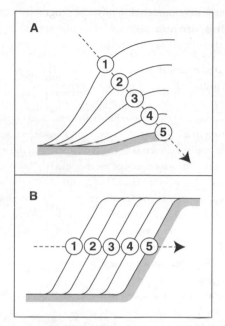

6. Early models of landscape evolution involving a 'cycle of erosion': (A) slope decline (the Davisian model) in temperate environments; and (B) slope retreat, more applicable in semi-arid environments. The numbers indicate successive stages in the evolution of hillslopes

'stage' that was paramount as so little was known about the effects of geomorphological processes on the underlying geological structure. Comparable evolutionary or developmental models were developed in other branches of physical geography. An American ecologist, Frederick E. Clements, coined the term 'climatic climax' for the terminal stage of a succession of plant communities. His role in biogeography was similar to that of Davis in geomorphology. In climatology, similar ideas were developed to explain patterns of climatic change, which are outlined later.

Although these developments in physical geography were moves towards a more focused and scientific approach, physical geography also played a major role in regional geography, which became the dominant theme within geography as a whole during the first half of the 20th century. In published regional geographies, a largely descriptive account was provided of the physical background to human occupation and use of different parts of the Earth's surface. Such accounts generally took the form of chapters on landforms, climate, vegetation, and soils.

Towards the end of the first half of the 20th century, therefore, physical geography was developing only slowly. The main area of progress was in the global classification of landforms, climates, vegetation, and the other phenomena making up the Earth's surface. The knowledge base was being broadened and organized. The characteristics and diversity of vegetation formations, of climate types, and of landform assemblages were being more fully catalogued. Although the classifications were often linked – types of climate to vegetation formations, for example – there was little attempt to understand the key Earth-surface processes that underpinned these phenomena. Earlier attempts to develop an integrated physical geography of the geo-ecosphere were being largely forgotten.

New directions

As interest in traditional regional geography waned around the mid-20th century, there was rapid growth and diversification within physical geography. This take-off point in the emergence of modern physical geography can be attributed to two main interrelated developments: first, the 'quantitative revolution' in geography as a whole, which brought an explicit emphasis on scientific method; and second, a 'process revolution' within physical geography itself, which brought greater understanding of the processes that produce the variable characteristics of the

Earth's surface. Both developments led physical geographers, once more, to seek methodological and substantive inspiration from cognate natural environmental sciences.

Earth-surface processes

The justification for focusing on 'process' rather than 'form' was that to explain spatial patterns on the Earth's surface and the dynamics of the geo-ecosphere, an understanding of processes and mechanisms is essential. It is not sufficient, for example, to propose models of the cycle of erosion as depicted in Figure 6 without understanding the processes that are affecting the slopes. Such ideas were truly revolutionary because they modified physical geography, root and branch, in fundamental ways. By investigating processes directly, physical geographers have demonstrated that many early inferences about the nature of the landscape were incorrect. Fluvial geomorphologists waded into the rivers to measure velocity and discharge; glacial geomorphologists dug tunnels in glaciers to observe the effects of erosion at the glacier bed; and desert geomorphologists used wind-tunnels to study the deposition of sand dunes experimentally. Description and classification of Earth-surface phenomena became increasingly replaced by the measurement, monitoring, analysis, and modelling of formative processes.

Despite holding a 'minority stake' within physical geography, climatology can be credited with having led the way in the process revolution. The effects of meteorological processes are clearly apparent in regular diurnal and seasonal weather patterns. Both the long-term average conditions and the short-lived extreme events that constitute climate are relatively easily observed. There was a long tradition of measuring and monitoring the meteorological elements – temperature, pressure, precipitation, and wind: indeed, national and international networks of meteorological stations were up and running over much of the Earth's surface. Climatologists were, moreover, familiar with the physical principles underlying meteorological processes, and they routinely used elementary statistics in summarizing their data. It was therefore less of a conceptual or practical leap for climatology to focus on processes. The result was explanations for climatic patterns based on the general circulation of the atmosphere, driven by energy from the Sun and incorporating smaller-scale circulation patterns, such as the mid-latitude depressions, and modified by regional and local effects, such as topography and surface characteristics. Today, the knowledge of such processes is being used as essential inputs to understanding the climatic changes that are occurring at the beginning of the 21st century (as discussed in Chapters 4 and 6).

Soil geography illustrates well the importance of processes to physical geography by linking the different types of soil to the different environments in which they form. Several soil-forming processes transform the unconsolidated material cover of the Earth's surface (the regolith) into productive soil. Each involves a set of physical, chemical, and/or biological transformations, and their effectiveness varies in different parts of the world. Leaching, for example, refers to the removal of soluble constituents from upper soil horizons: it requires downward percolation of water through the soil and commonly occurs where rainfall exceeds evaporation. Eluviation, the mobilization of clay in the upper horizons and its redeposition in the subsoil, is best developed in humid climates with a dry season. In contrast, upward movement of soluble salts (salinization) and their accumulation in the soil profile is characteristic of arid and semi-arid lands where there is insufficient rainfall to wash them out. The salts are drawn up in groundwater by capillary action and left behind as the water evaporates. Little understanding of the variability of the world's soils is possible without knowledge of these processes going on in the soil. Furthermore, the greater our understanding of soil-forming processes, the more likely it is that soil management and conservation will be effective, soil productivity maintained, and soil degradation minimized.

Systems

Widespread adoption of a 'systems approach' developed out of the 'process revolution'. In general, a system can be defined as a set of objects together with their interrelationships. A systems approach downplays the individual objects and places the interrelationships between them at centre stage. Understanding the interrelationships within a system naturally requires consideration not only of structural connections but also how they are functionally related: hence processes are central to understanding any system. Another important aspect of a systems approach is the blurring of boundaries – because each system is connected to the next. Many different types of Earth-surface systems can be recognized within the geo-ecosphere and its component spheres, and they are all interconnected to a greater or lesser degree in the landscape. A systems approach is therefore applicable throughout physical geography at a wide range of scales, from the explanation of particular phenomena in the landscape to understanding the dynamics of how the whole geo-ecosphere operates. Its adoption can also be seen as heralding a return to a more integrated physical geography.

One of the first and most influential applications of the systems approach in physical geography was the adoption of the ecosystem concept by biogeographers. An ecosystem includes a set of organisms together with their interactions and environmental relationships. In coining the term and developing the concept, ecologists focused on energy flow and mineral cycling between green plants (producers), animals (consumers), and micro-organisms (decomposers). In the physical geographical context, vegetation formations of the Earth, for example, are viewed as parts of geo-ecosystems with different levels of energy inputs that maintain different levels of productivity and are capable of sustaining different types and numbers of animals, including humans.

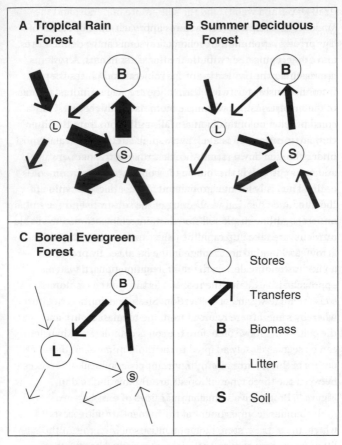

7. **Mineral cycling in three of the world's major forested geo-ecosystems: (A) tropical rainforest; (B) deciduous summer forest; and (C) boreal evergreen forest**

A simple comparison of three of the world's major geo-ecosystems (tropical rainforest, summer deciduous forest, and boreal coniferous forest) in terms of a three-compartment systems model of mineral cycling is shown in Figure 7. The three compartments (circles) – the biomass (mostly vegetation), the litter (mostly dead leaves and wood), and the soil – store minerals; and the size of each circle represents the quantity of minerals stored. The arrows represent the cycling of minerals between the stores, and other inputs and outputs; the thickness of each arrow representing the proportion of the minerals that is transferred from the source store each year. In the tropical rainforest (A) most of the minerals are stored in the biomass (mostly trees). Only a small proportion of this biomass falls as litter to form the litter compartment, which is a relatively small circle also because most litter is broken down rapidly (thick arrow) in the temperature and moisture conditions on the forest floor. The environment is optimal not only for tree growth but also for decomposition of the litter and chemical weathering of the substrate and the soil minerals. Although the soil store is of intermediate size, available nutrients are taken up rapidly by the tree roots (another thick arrow) and tend to accumulate in the biomass. By providing an understanding of how the rainforest functions, a systems approach also explains the reduced fertility of the rainforest soils after a few years of cultivation. Slash-and-burn agriculture, whereby minerals are released from the biomass by fire and the cultivators move on before the soil is completely exhausted, can be seen as closely adapted to the rapid mineral cycling that occurs in this environment. Important geographical differences between the three types of forests are also highlighted in Figure 7. Progressively smaller quantities of minerals are stored in the biomass compartment of the higher-latitude forests which, in contrast, store larger quantities of minerals in the litter compartment. The latter is largest in the boreal forest (C), where a relatively small proportion of the minerals is transferred to the soil and made available to the trees because litter decomposition is slowest there, especially under cold winter conditions. In contrast, the soil compartment of the summer deciduous forest (B) is largest because both decomposition of the litter and uptake of minerals by the roots are at intermediate levels.

Early applications of the systems approach within physical geography consolidated the shift of emphasis from Earth-surface forms to processes and from long-term development of the landscape to how it is maintained in a state of balance (steady-state equilibrium) over the short term. It was soon realized, however, that landscape systems are more complex than this and are rarely stable for long. Instability may be caused by the internal dynamics of the system or by natural or human-induced disturbance. In the case of soils, for example, the natural leaching process may gradually exhaust the nutrients and lead to a decline in the vegetation cover and hence soil erosion. Poor agricultural practices have similar disruptive effects, while extreme natural events such as floods and hurricanes may erode the soil more abruptly. Landscape systems differ in their sensitivity and resilience to disturbance and in the thresholds that have to be overcome before a system is tipped from one relatively stable state to another. These concepts are important at the frontiers of physical geography today. All aspects of the natural landscape – whether reference is being made to the surface landforms, the vegetation and soil cover, or the climatic envelope – are subjected to disturbance and change.

Long-term environmental change

Physical geographers now regard the study of past, present, and future environmental change as very important. The environmental change theme rose to prominence towards the end of the 20th century but the seeds were sown earlier. It may be traced back to members of the Helvetic Society in Switzerland in the first half of the 19th century. The president of that society, Louis Agassiz, who published the book *Études sur les Glaciers* in 1840, was both an early convert and a particularly influential advocate of their ideas. They recognized that many of the erosional and depositional features apparent on the forelands of Alpine glaciers were similar to those in the landscapes of the Alpine foothills, the North German Plain, and the British Isles, far away from any present-day glaciers. The evidence included signs of the power of glacial erosion, (ranging from bedrock surfaces polished by abrasion to deeply excavated 'U'-shaped valleys), erratic boulders transported by glaciers for long distances from their source areas, and moraines (ridges or mounds of unsorted sediment deposited by glaciers). This evidence also illustrates the fact that many landscapes cannot be explained by the processes currently acting upon them.

Major conclusions of the early 'glacialists' were that extensive glaciers and ice sheets once covered a much larger proportion of the Earth than today, and that the Earth's environments had recently been affected by an 'Ice Age'. Subsequently, it was revealed that there had been more than one 'glacial' episode, when global mean annual temperature was at least 10 degrees Centigrade lower than today; that these 'glacials' were separated by 'interglacials' during which climatic conditions were much like those of today; that global sea level varied by over 100 metres due to the abstraction and release of water into the oceans as the ice sheets waxed and waned; and that all the components of the geo-ecosphere were profoundly affected, including the tropical, arid, and warm temperate regions not directly affected by glacier ice. These deductions provided an alternative explanation of features previously attributed to Noah's flood. They also represent the first steps in our modern understanding of Quaternary environmental change, in which physical geographers play an important role. The Quaternary is the geological term for the most recent major period of Earth's history: it includes the present day and has lasted for more than two million years.

Of the many advances in understanding Quaternary environmental change, two can be regarded as particularly important. First, in the 1930s, Milutin Milankovitch, a Serbian applied mathematician, developed the 'astronomical theory', which successfully explains the regular climatic changes associated with glacials and interglacials – the 'pulsebeat' of the Ice Age (see box).

The astronomical theory of climatic change

This mathematical theory, also known as the 'Milankovitch theory', accounts for the regular, long-term variations of climate that produced glacials and interglacials over the Quaternary period. The theory was disputed during Milutin Milankovitch's lifetime but was later fully tested and is now widely accepted. It predicts the quantity and distribution of solar radiation received at the Earth's surface in response to regular variations in the distance of the Earth from the sun. This depends on three so-called orbital parameters: the *precession* of the equinoxes, which varies with a periodicity of about 21,000 years; the *obliquity* of the ecliptic, which varies with a periodicity of about 41,000 years; and the *eccentricity* of the orbit, the periodicity of which is about 100,000 years. These orbital variations can be envisaged, respectively, as measures of the 'wobble' and 'tilt' of the Earth about its axis and the 'stretch' of the Earth's orbit (the extent that the elliptical orbit departs from a circle). The three orbital parameters combine to determine the pattern and timing of glacials and interglacials: the former correspond with times of minimum solar radiation receipt; the intervening intervals are the interglacials.

The first decisive test of the astronomical theory utilized data from marine sediment cores, which retrieved the material that had slowly accumulated on the deep ocean floor during successive glacials and interglacials. Specifically, the oxygen

isotope ratio of shells of microscopic plankton (which
reflects the volume of water in the oceans and ice sheets) was
compared with the predictions of the theory using statistical
techniques, and a close match was found. Since then, further
successful tests have been made on, for example, coral
reef sequences from Barbados (which reflect the sea-level
variations that accompanied the growth and shrinkage of
the ice sheets) and ice cores from Antarctica and Greenland
(reflecting changes in properties of the atmosphere).

The second advance was the reconstruction from marine
sediments of continuous records of the actual climatic changes
that affected the Earth over this same time interval. Information
about environmental change is contained in the remains of
microscopic organisms that have steadily accumulated, often
without major breaks and mostly unaffected by subsequent
erosional events, on the ocean floor. This source of information
could not be tapped, however, before the technology had been
invented to retrieve successfully sediment cores from the deep
oceans. For most of the first half of the 20th century, the strong
belief, based on terrestrial evidence from Europe and elsewhere,
was that there had been only four glaciations during the
Quaternary. It is now known that there were more than 10 times
this number of glacial episodes within the current Ice Age. The
marine evidence, which allowed Milankovitch's theory to be tested
definitively, today provides a more-or-less complete temporal
framework for the shorter, often discontinuous records that are
available from the terrestrial Earth surface.

The effect of these advances was to draw physical geographers,
along with many other natural environmental scientists, into
reconstructing the pattern, timing, and effects of these major
climatic events. In the terrestrial context, there was first an
emphasis on reconstructing vegetation change from mires based

on pollen analysis, which was later supplemented by information from a wider range of 'natural archives' including, for example, lake sediments, loess (wind-deposited silt), ice cores, speleothems (precipitated cave sediments), tree rings, and corals, which together provide data on past changes in all the component spheres of the geo-ecosphere.

With continuing development of new techniques for accurate dating and precise measurement of 'proxy' environmental data from such archives, it has been possible to reconstruct past environmental changes in great detail. Ice cores, for example, obtained from drilling through the Greenland and Antarctic ice sheets, have revealed changes in atmospheric composition through the last several glacial-interglacial cycles, have demonstrated the existence of abrupt environmental changes over relatively short timescales (temperature increases of 10 degrees Centigrade have occurred within decades), and have shed light on the complexities of the interactions within the land-ocean-atmosphere system.

With increasing realization of the importance of past and present human impacts on the geo-ecosphere, there has also been a new emphasis on the short-term climatic changes of the Late Glacial and the Present Interglacial or Holocene (the last 11,500 years) and their effects. For example, decadal- to millennial-scale climatic variations underlie the Holocene glacier variations illustrated in Figure 8. Episodes of glacier expansion (neoglacial events) have been reconstructed from the record of suspended sediment deposition in stream-bank mires and lakes downstream of glaciers in the mountains of Jotunheimen, southern Norway. Glacier expansion and contraction resulted in variations in the production of fine sediment (especially silt) by glacier erosion. These variations are in turn reflected in the thickness and composition of the sediment layers retrieved by excavating the mires and coring the lake sediments. Radiocarbon dating of organic material provides the detailed timescale.

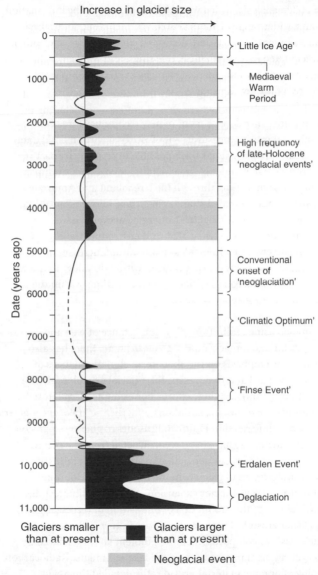

8. Holocene glacier and climatic variations in Jotunheimen, Norway, reconstructed from the sedimentary record in lakes and mires

The neoglacial events were a response to either a decrease in summer temperatures, which leads to enhanced melting of the glacier ice, or an increase in winter (snow) accumulation, which leads to glacier growth. Glacier variations in effect filter out the year-to-year climatic variability and have revealed that the climate often differed appreciably from that of today. Some of the features of the record are named in Figure 8, including the likely absence of glaciers during the so-called 'climatic optimum' of the mid-Holocene, around 7,000 years ago (characterized by temperatures 2–3 degrees Centigrade warmer than today); the re-growth of glaciers (conventional neoglaciation) between about 6,000 and 5,000 years ago; and the 'Little Ice Age' of the last 500 years (with temperatures 1–2 degrees Centigrade colder than today).

There are at least two wider implications of both the long-term and the short-term record of past climates that provide strong reasons for physical geographers to continue to investigate the environments of the past. First, similar natural variations in climate will almost certainly continue to occur in the future. It is therefore vital to understand the natural background onto which anthropogenic global warming is being superimposed. Second, climatic variations are the drivers of changes in so many other aspects of the geo-ecosphere that are central to physical geography and so important to humanity.

Human impacts: from Holocene to Anthropocene

As the 20th century drew to a close, human impacts on the geo-ecosphere and, in particular, on global climate became a priority, both for science and for society. The extent and intensity of the impacts, and rates of change in the biosphere, pedosphere, hydrosphere, and toposphere became significant from the mid-Holocene, around 5,000 years ago. Since then, from Neolithic times onwards, a succession of increasingly advanced technologies has enabled the increasing exploitation of the Earth's resources with accompanying intentional and non-intentional environmental effects. Deforestation, soil degradation, and hillslope erosion have had a particularly long history, and most early civilizations, from Mesopotamia to Middle America, experienced major impacts of these processes. Physical geographers recognized the seriousness of many of these impacts over a century ago – as demonstrated, for example, by George Perkins Marsh in *Man and Nature; or Physical Geography as Modified by Human Action*, published in 1864.

It is only since the Industrial Revolution, with the extensive burning of fossil fuels, that there has been a significant human impact on global climate. Although the first quantitative estimate of the enhanced greenhouse effect produced by the release of carbon dioxide was made by a Swedish scientist, Svante Arrhenius, as early as 1896, it is only as we enter the 21st century that the scale of the effects of such anthropogenic greenhouse-gas pollution on the atmosphere is beginning clearly to exceed the scale of the effects of natural climatic variability. In recognition of the general prevalence of human impacts and the unprecedented rates of change being induced throughout the geo-ecosphere, the term 'Anthropocene' has been coined for the most recent 200 years or so of Earth's history (see box).

The environmental change theme in physical geography has, therefore, been re-invigorated and has itself changed as future climates due to enhanced global warming have become not only a pervasive influence on the research agenda in the natural environmental sciences but also a topic of everyday conversation and an increasingly potent force on the political agenda. Relatively long-term environmental change is particularly important for the current focus on carbon emissions as it provides a test bed for understanding the natural carbon balance of the land-ocean-atmosphere system. The shorter-term changes of the Holocene provide insights into the contemporary natural background variability onto which the

The Anthropocene

The term 'Anthropocene' was proposed by Paul J. Crutzen and Eugene Stoermer in the year 2000 to define a new geological epoch characterized by the dominance of human impacts on the geo-ecology of the Earth. It covers approximately the last 200 years of the Holocene. During the Anthropocene, the world's population has increased to over 6 billion people, and the scale of human exploitation of the Earth's resources is unprecedented. At least 50% of the land surface of the Earth has been transformed by human actions. More than 50% of all accessible fresh water is now used by humans. Around 20% of mammal species, 10% of bird species, and 5% of fish species are currently threatened with extinction as a result of human activities. The artificial nitrogen added annually to soils as fertilizer now exceeds the amount fixed naturally in soils. Sulphur dioxide emissions to the atmosphere from fossil fuel burning and tropical forest fires are now double those emitted naturally. Atmospheric carbon dioxide and methane (two important greenhouse gases) have increased by about 30% and about 150%, respectively, during the last 200 years.

rapid changes of the Anthropocene are being superimposed. The immediate challenge for physical geography is how best to contribute to accurate knowledge and improved understanding of the science of environmental change – including the human impact – so that informed political decisions can be made, not only about how to achieve an acceptable carbon balance, but also about how to deal with the many other human 'footprints' being made on the Earth's surface. Environmental changes of the Anthropocene will therefore reappear later in this chapter, and many times later in the book, in the context of geography as a whole.

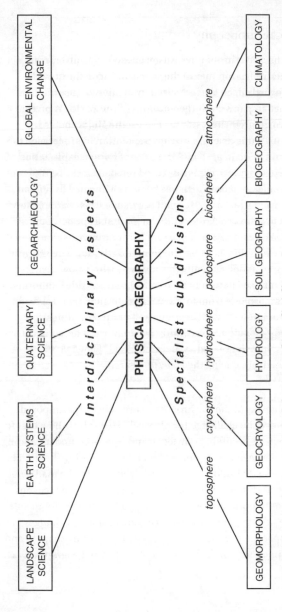

9. Physical geography: its specialist subdivisions and interdisciplinary aspects

Physical geography today

A key characteristic of physical geography is its diverse subject matter. This has, on the one hand, encouraged the growth of specialisms and, on the other, required physical geographers to look outwards towards other subjects. A simplified representation of the current structure of physical geography is shown in Figure 9. In this diagram, the specialisms are complemented by their links with other disciplines. The main specialisms, such as geomorphology and climatology, correspond with the different spheres and are summarized in the box. The interdisciplinary linkages allow physical geographers to work, for example, with geologists, biologists, or archaeologists; and although not interdisciplinary, collaboration with human geographers might be included here. At the centre of the diagram, integrated physical geography involves the study of whole geo-ecosystems: more than one element of the landscape – not just landforms, vegetation, soil, or climate – and the scale may vary. Examples of the landscape units that provide the basis for integrated physical geography include hillslopes, river drainage basins, lake catchments, cities, mountain regions, and the whole Earth.

Specialisms of physical geography

Six main specialisms can be recognized, each covering a major component of the geo-ecosphere. All involve study of the Earth's surface today and in the past and how it is likely to change in the future, both naturally and with increasing human impacts.

Geomorphology focuses on landforms and their formative processes, especially the Earth-surface processes involving weathering and erosion, transport and deposition by water, wind, and ice. 'Mega-geomorphology' includes global-scale landforms, such as mountain ranges.

Hydrology studies water in the land-atmosphere-ocean system, especially the processes of the hydrological cycle (such as evaporation, precipitation, and run-off) and their applications (such as in floods and droughts). It includes the study of groundwater and water in the cryosphere.

Climatology investigates not only the average condition of the atmosphere but also the range, frequency, and causes of atmospheric events. It includes mass and energy exchange between the Earth and atmosphere, the general circulation of the atmosphere, and the regional and local atmospheric circulations that account for weather patterns.

Biogeography studies all aspects of the distribution of life – the 'living world' of plants, animals, micro-organisms, and whole ecosystems. Major interests include the vegetation formations of the Earth, such as tropical rainforest, savanna grassland, and tundra, and their biodiversity.

Soil geography, sometimes regarded as part of biogeography, focuses on variation in the soil cover of the Earth, especially on differences in the development and degradation of soils.

Geocryology, overlapping in part with hydrology and also with geomorphology, is the study of the Earth's snow, ice, and frozen ground, including glaciers and ice sheets, perennially frozen ground (permafrost), and seasonally frozen ground.

The interdisciplinary contribution

The breadth of physical geography means that there are many actual and potential mutually beneficial interactions with established cognate sciences. Some of these links have developed into well-established interdisciplinary research fields, such as Quaternary science and geo-archaeology. In Quaternary science, physical geographers work in teams with Quaternary geologists, palaeo-climatologists, palaeo-ecologists, and others to reconstruct past environmental changes. In geo-archaeology, physical geographers join with archaeologists to focus on the human past. This field, closely related to environmental archaeology, brings the approaches, methods, and concepts of the natural environmental sciences to bear on the interpretation of archaeological sites. The physical geographer's contribution is particularly apparent in understanding the 'natural site formation processes' (such as erosion and deposition) and their interactions with the 'cultural site formation processes' (such as site construction, usage, and modification).

Physical geographers also contribute to several emerging interdisciplinary sciences related to human environmental impacts and their future management, such as sustainability science and Earth system science. The former focuses on the ways and means of maintaining and enhancing the Earth's productive functions in the face of destructive exploitation of resources; the latter focuses on understanding the total Earth system without necessarily understanding all the individual processes and mechanisms. Physical geography contributes its spatial perspective, its holistic tradition, and its concern with the human dimension. It thereby makes an important input to the solution of some of the big scientific questions and associated practical problems facing humanity today.

How useful is physical geography?

Physical geography is most conspicuously useful in relation to the many problems that people have to face in the environment. Such problems occur at all scales and are illustrated here by example. (Many more examples are given later in the book.) First, geomorphological mapping and terrain evaluation for planning the route of major roads provides a good example of a specialist application at the local scale. This involves mapping topographic variations and substrate characteristics, including assessment of drainage conditions and slope stability. A routeway with suitable gradient and stability can then be identified and, if necessary, engineering to ameliorate conditions and reduce hazards can be carried out prior to construction work. Specific additional problems are associated with particular environments, such as the thawing of permafrost under roads in periglacial regions and the salt weathering of concrete in arid lands.

Another example relates to the impacts of climate change in the marginal environments that are most susceptible. In the African Sahel, enhanced global warming with a slight increase in the frequency or severity of droughts is likely to be far more disastrous than the same effect in an area with greater moisture availability. The major issue of desertification is strongly related to such changes, though human factors are also centrally involved. Similarly, the sea-level rise in the order of 50 centimetres that may occur over the next few decades is of greatest relevance on low coasts (such as deltas or coral islands), especially in developing countries, where the people are less able to adapt in the face of such natural hazards. The breadth of physical geography ensures it is useful in diverse ways, but perhaps it is most useful when physical geographers apply their knowledge, understanding, and techniques in collaboration with other scientists and with human geographers.

The roles of physical geography

It is possible, therefore, to view physical geography as playing three fundamental roles. First, it contributes its distinctive spatial perspective to gaining knowledge and understanding of each of the component parts of the landscape. This role is clearly the driving force of its specialisms. Second, it explores the interconnections between those components, relying upon a more holistic approach to the landscape than any other science. This is perhaps the core role of physical geography as an entity. Third, it has the interface between the natural environment and people as a central concern, which consolidates its role as part of geography as a whole.

Integrated physical geography, despite a long tradition dating back to Alexander von Humboldt, is currently the least developed of these roles. It nevertheless has the potential to bring together the various specialisms and provide a basis for developing further links with interdisciplinary science and with human geography. This line of argument suggests that integrated physical geography could be a focus around which to build a new physical geography defined as:

> that branch of geography concerned with (a) identifying, describing and analysing the distribution of biogeochemical elements of the environment; (b) interpreting environmental systems at all scales, both spatial and temporal, at the interface between atmosphere, biosphere, lithosphere and society; and (c) determining the resilience of such systems in response to perturbations, including human activities.

O. Slaymaker and T. Spencer, *Physical Geography and Global Environmental Change* (1998)

Chapter 3
The human dimension: people in their places

Human geography has moved through a series of changes in approach and content to a point where it is now extremely diverse. These qualities raise challenges of focus and definition but also introduce a rich range of topics, innovations, and subject matter. The study of human geography once had a clear and unambiguous meaning. It was concerned, in particular, with the ways in which people occupied the surface of the Earth: the patterns of settlement that emerged, the human landscapes that evolved, the movements of human populations that occurred, and the 'order' that became apparent. When the question arose of explanation to clarify issues such as why cities were located in particular places or why there were high concentrations of population in some parts of the world and great voids in others, the answers were usually sought firstly in the natural environments and secondly in history. These initially served the purpose but became constraints upon human geography and its relative paucity of good theory. Whereas the subject matter and the questions were valid, the sources of explanation were far from complete and excluded huge areas that have now become central concerns of human geographers.

Changing approaches: rocking the traditional boat

One of the early grand theories, environmental determinism, exemplified the emphasis on the natural environment in explaining human actions: the latter were presumed to flow from the former. It was suggested, for example, that people living in hot climates tended to be lazy and promiscuous and that crimes of violence were higher in hotter parts of countries, such as the 'violent south' in the United States. The theory was by no means bereft of virtue but it did produce many blind alleys and unverifiable explanations. Regional geography, or studies of particular territories on the Earth's surface, initially followed a similar rather mechanical pathway, but eventually led to a stronger development of the proper understanding of human activities, paying more attention to history and culture. Something of the mindset and relative longevity of these early attempts at understanding causation in human geography is revealed in the British geographer K. G. T. Clark's paper published in 1950. He argued that human geography was making very limited headway because of the widespread assumption that physical geography was *the* necessary basis for understanding the human phenomena in question. Since that time, the pace of change has been remarkable.

There are several defining characteristics of these changes that included a major move away from the natural environment as a primary or major source of explanation; an eventual radical qualification of objective reality as a prime focus of study and a greater awareness of the fact that different people perceived and experienced the world in different ways; and a search for sources of theory and inspiration well outside the traditional boundaries of human geography. One outcome has been an increasingly eclectic human geography, vulnerable to the critique that 'anything goes', described as being in wild pursuit of its identity and as undergoing a feverish bout of deconstruction and reconstitution.

Such changes are experienced by many academic disciplines but seem at times to have reached extremes in human geography. These changes are often called paradigm shifts, in which one

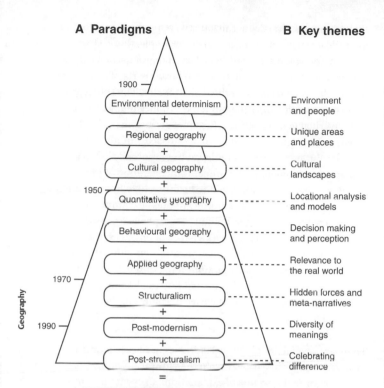

A Paradigms

B Key themes

1900

Environmental determinism ---- Environment and people

\+

Regional geography ---- Unique areas and places

\+

Cultural geography ---- Cultural landscapes

1950

Quantitative geography ---- Locational analysis and models

\+

Behavioural geography ---- Decision making and perception

\+

Applied geography ---- Relevance to the real world

1970

\+

Structuralism ---- Hidden forces and meta-narratives

1990

Post-modernism ---- Diversity of meanings

\+

Post-structuralism ---- Celebrating difference

=

PRESENT-DAY HUMAN GEOGRAPHY

Geography

10. Paradigms of human geography: (A) the historical sequence of paradigms that has left its legacy to present-day human geography; and (B) the key themes that they represent. An approximate timescale for the take-up of each paradigm and the broadening base of human geography are also indicated

prevailing and influential set of ideas is replaced by another (Figure 10). The old paradigms rarely disappear altogether but leave a legacy capable of resurfacing in some new and amended form. As indicated, the first expressions of cause and effect in human geography were strongly influenced by the need to relate people and settlement to the natural environments that they occupied. Any form of explanation was couched in terms such as mountainous areas produce dispersed forms of settlement and the plains foster nucleation; desert areas lead to nomadic economies whereas fertile lowlands produce dense populations and large clusters. Generalizations of this kind offered some credible insights but exceptions abounded. As human geographers became more aware of the discrepancies, they turned towards more historical and cultural sources of explanation.

The broader way of thinking that evolved in regional studies recognized that people could modify landscapes: cultural traditions could persist and strengthen over time; and there were human conditions that countered and often over-rode environmental imperatives. In other words, landscape was a human record as well as a physical artefact. Much of this earlier human geography was concerned with *regionalism* and the study of places. In 1939, Richard Hartshorne, a leading American geographer, published his *Nature of Geography*, in which areal differentiation, or the study of areas or regions on the Earth's surface and their causally related differences, was proposed as the key quality of what geographers did. This rendition of geography, with its focus on the unique blend of factors that produced distinctive regions, reigned supreme throughout the middle decades of the 20th century.

This paradigm came under intense criticism in the 1960s, and urban geographers were among the main advocates of the need for change. The established approaches for studying the geography of towns and cities were accused of being mainly descriptive, lacking in good measurement techniques, and failing to develop sound theories. The remedy proposed was spatial science that applied scientific methods to geographical phenomena.

The rise and fall of spatial science

As the new paradigm of spatial science became translated into human geography, it had several distinctive characteristics (see box). The new paradigm led to the 'Age of Models'. Long-extant versions, such as Von Thünen's land-use zones of 1826, developed by a German landowner and land economist, and the German geographer Walther Christaller's central place theory of the 1930s, came into prominence. Both were concerned with the ways in which land-use and settlement would develop over a 'model' uniform surface. For Von Thunen, the key to rural land-use was the fact that land nearer to the farm holding would be worked more intensively and produce crops such as horticulture, whereas more distant fields would be used extensively for pastoral farming. Distance from the farm was a cost and the economics of land-use could be best managed in this way.

Spatial analysis and the quantitative revolution

This approach developed from the 1950s and was designed to make human geography (and geography as a whole) more scientific. The approach emphasized the need to turn away from the unique aspects of the Earth's surface – as embodied in regional geography – and to follow the philosophy of science in seeking generalizations that could be verified. As these principles were imported into spatial analysis, some key features were:

- An interest in patterns and shapes or the geometry of space.

- The use of representative samples.

- The use of measurement, numerical methods, and statistics.

- The development of testable hypotheses, models, and theories.

- A search for models and algorithms with predictive power to allow, for example, optimal locations to be identified and spatial change over time to be analysed.

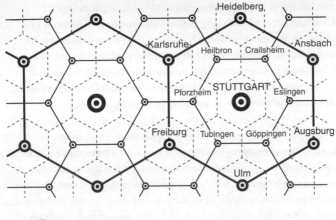

11. Walther Christaller's central place model, characterized by repeated hierarchical hexagonal patterns: towards the right, actual settlements from southern Germany have been mapped onto the model

Christaller argued that consumers of goods in a market town would respond to the constraints of distance by travelling to the nearest centre that offered that particular commodity. His nearest centre hypothesis was a simple and mechanical way of explaining and describing human behaviour. Given this hypothesis, a specific pattern of market towns would develop on a uniform plain, and Figure 11 shows the classic form of the Christaller model with its system of a hierarchy of central places and nested market areas in a hexagonal pattern. Here just three orders of central places and market areas are shown, with an attempt to fit the model to southern Germany using Stuttgart as the high-order centre. The essence of this model was that it assumed a uniform plain (often referred to as a billiard table surface) with equal accessibility from all directions; distance therefore became the key factor in explaining the location of market towns, the definition of their market areas, and the pattern of consumer behaviour. For example, towns of the same rank were the same distance apart and their market areas were the same size; consumers would respond to distance by using the nearest centre. As the model conditions do not exist, attempts to fit the model to the real world, as in the example shown, can only be approximations.

The models became increasingly sophisticated and were accompanied by many attempts to measure geographical patterns and processes: training in statistics and quantitative methods became a *sine qua non* for geography students. This was a strong and vibrant new paradigm that relegated older approaches to the background. Its many methods included multivariate analyses applied to urban studies to handle large sets of information, often from census data, predictive models for economic geography to examine projected patterns of growth, and a whole raft of diffusion and simulation studies in fields such as migration, spread of innovations, and diseases, which allowed best estimates of the ways in which flows would occur over space and time. Statistical data sources, especially those provided by census small area statistics and other official statistics, supplemented by social surveys, came to the fore.

By the late 1970s, however, doubts about the value of this paradigm were emerging. The methods had become far more sophisticated but the outcomes were still descriptive; there were plenty of models but not so many well-founded theories. With its emphasis on the geometry of space, spatial analysis produced theories based on oversimplified assumptions of human decision-making. It employed the concept of 'economic man' who was rational, had perfect knowledge, optimized opportunities, and minimized costs. Economic man, for example, would obtain services from the nearest available location, assess all the options when changing place of residence, and locate a business where all necessary conditions were met. Human geographers began to question these assumptions as being far removed from the reality of people's behaviour. In reality, people often behave in suboptimal ways, with much less than full information and perhaps influenced by emotional values.

From this point, spatial analysis was to lose its predominance and another process of relegation was in train, though the impact of the changes varied in different parts of the world. By the early 1970s, new perspectives began to dominate the practice of human geography. Humanistic and cultural approaches in particular discarded the scientific perspectives of spatial analysis and moved away from its emphasis on measurement and generalization. The new focus was on qualitative meanings and values and on the diversity of human behaviour. There was evidence for the inadequacy of central place theory, for example, as many consumers were shown to ignore its precepts and shop at a variety of destinations. The sophisticated models of spatial analysis bore little relation to the real lives of people and their attachments to specific places; it was time to assert the diversity of human behaviour. For many, this was a welcome and positive shift, but others argued that geography had turned its back on quantitative spatial data analysis just as many other disciplines, such as economics, came to recognize its importance.

First reactions: humanistic and structural approaches

There were initially two main points of departure from the focus on spatial analysis. Humanistic geography asserted the centrality of people and focused on the meanings of place rather than the geometry of space. The 'image' entered the lexicon of geography with ideas such as perceived space, mental maps, and 'irrational' behaviour. It was evident that different people perceived places in different ways. There were studies, for example, of the meanings of 'wilderness'. At one extreme, there were those who defined wilderness as remote areas devoid of population; at another were those who regarded well-used country parks as fitting the description. Again, the mental map of a neighbourhood held by an elderly person with constraints on his or her mobility was very different to that of a younger individual. Many people would use their nearest shopping centre, but some were prepared to travel further afield because they had the personal mobility and preferred a different venue. Geographers turned to new sources of information such as works of art and fictional writing for their insights into landscape and place. There were studies of individual painters and specific works that could be used to examine the changing nature of landscape and landscape gardening over time. Works of fiction, such as *The Grapes of Wrath* and many of the novels of Thomas Hardy, were used as sources to throw light on contemporary society. This 'subjective' reaction matured into the *new cultural geography*, and that warrants further consideration.

Firstly, however, we can identify the second reaction to spatial analysis, which can be described as structuralism (see box). Basically structuralism offered a grand theory that explained both human behaviour and its societal outcomes. A central idea is that there are powerful forces within society that condition the kinds of lifestyles that can be followed. Capitalism provided one such powerful and conditioning force, and it was argued that the growing problems of disparities of wealth and quality of life within cities and between societies stemmed from its influence. David Harvey, a British geographer working in the USA and a leading Marxist scholar, argued that there was a clear disparity between the sophisticated theoretical and methodological frameworks of spatial analysis and geographers' ability to say anything really meaningful about events as they unfolded around them. In other words, geographers were not grasping the significance of these major structural forces and were only dealing with the surface manifestations of a deeper process. Structuralists turned to Marxist theory and the notion of hidden structures that had a

Early reactions to spatial analysis

A widely held view had developed by the 1970s that although spatial analysis had given human geography a sound scientific methodology, it remained largely descriptive and had failed to develop good theories. Many of the assumptions upon which it rested were unrealistic and bore little or no relation to the diversity and complexity that actually existed in the real world.

Humanistic geography sought to reassert the importance of people and raise them from their status as 'pale entrepreneurial figures'. It introduced a new focus on subjective values and qualitative meanings that affected people's behaviour. It proposed the importance of the image and the perceptions of geographical space that people held as mental maps, shaped by their circumstances and experiences.

Structuralism suggested that the sources of explanation were found in the hidden structures of empowerment and control that underlay different types of society. Marxism was one such structural theory that related distributions of wealth and poverty to the workings of a capitalist society. Spatial outcomes such as areas of poverty in cities and other underdeveloped regions could be understood in these terms.

compelling, even deterministic influence on human activities. Working from this perspective, Harvey argued that the theory had a great deal to offer geography, but had to be modified and rethought to include the fundamental concepts of place and space. Structuralism, for example, could be used to explain divisions within society between rich and poor, but the concentrations of the poor in specific areas such as the inner city were the product of investors who discriminated against those areas. Red-lined

Reactions to grand theory

Postmodernism is perhaps best known as a style of art and architecture, in particular, that is characterized by its diversity of forms and lack of uniformity. Some of the more spectacular buildings in Paris, such as Les Halles and parts of La Défense, can be put in this category. It also has more general meaning as a movement against grand theory, or 'meta-narratives', and an assertion of the importance of differences and pluralities. It is this aspect that most affected human geography.

Post-structuralism is a position that rests heavily on the 'French connection' or the highly influential succession of French critical theorists who had major influence on human geography from about the early 1970s. Again, there is a strong opposition to grand theory but a major interest in language, signs, and the interpretation of texts. This has been described as a questioning of the relationships between situations and the ways in which they are represented.

Discourse analysis is an approach that enables us to reveal the hidden motivations behind a text or behind the method that has been used to interpret it. The search is less for an answer to a specific problem than an understanding of the conditions that underlie it and the assumptions on which it is based. Discourse analysis is often seen as a product of postmodernism because of the latter's rejection of the idea of a general belief system and its own view of a world that is inherently fragmented and heterogeneous.

neighbourhoods were starved of mortgage funds because of investment decisions that had a spatial frame of reference. On a wider scale, regional disparities in wealth and development could be explained by the market-driven decisions of corporate business.

Interest in Marxism waned, largely perhaps because of its historical foundations and also because of the demise of the 'socialist state'. In the debate between structure (deeper forces) and agency (individual decision-makers), the diversity of the latter was to gain favour over the hegemony of the former. The debate had opened the receptive doors of human geography ever wider; social theory and later 'critical theory' gained wider acceptance. As post-structuralism and postmodernism (see box) succeeded the structural debate, other theorists became the influential forces. There was a major expansion of the influence of a number of French intellectuals, such as Louis Althusser, the social theorist; Roland Barthes, the cultural critic; Jacques Derrida, the linguistic philosopher; and Michel Foucault, the historian. The kind of theories they produced dismissed meta-narratives or grand theory and focused on differences and on the multiple meanings of language and text.

This pattern of derivative ideas and theories, or turning to literature outside the mainstream of human geography, was not new. Park and Burgess who introduced the influential Chicago model for city structure and growth were social ecologists; Kevin Lynch, who conducted the early work on images in the city, was an architect. But the new vogue was for ideas, re-interpretations, and theories rather than for evidence-based research.

Contemporary human geography

Contemporary human geography has evolved from this pattern of changing paradigms and shifting priorities. It has moved from a simple and straightforward analysis of the relationships between people, settlement, and environment into a study of far more diverse and complex relationships. One can argue that the thread of space, place, and environment is still there, at least for many of the lines of study, but its nature is now very different. The preface to a study of the spaces of postmodernity offered a definition that captures the recent changes:

Human geography is that part of social theory concerned to explain the spatial patterns and processes that enable and constrain the structures and actions of everyday life. It provides an account of the ways in which complex socio-cultural, economic and political processes act through time and space.

M. J. Dear and S. Flusty, *Spaces of Postmodernity* (2002)

This definition is a long way removed from older ideas of people, settlement, and environment, but the cost is a diffuseness and diversity of positions in which a common core of endeavour is difficult to discern. Postmodern approaches emphasizing the heterogeneity of human existence, the differences and diversity within human populations, and a plurality of human geographies has emerged. Among the things that many human geographers seem, at least for the moment, to have lost along the way is the strong empirical tradition of evidence-based research, the science base, regional studies, and the analysis of interactions between people and their natural environment. Eric Sheppard, a Canadian geographer, speaks of the divergence between spatial analysis, with its components of quantification, spatial patterns, empiricist science, and general theory, and the new social theory, with its combinations of political economy and qualitative theorizing.

Against this background of argument and counter-argument, can we recognize a number of distinctive strands or find paths for reconciliation? There are some summary statements that can be made. Human geography can now safely be described as a loose confederation of approaches and ideas that often conflict with one another. Some of these approaches, such as the people/environment debate and the cartographic tradition, go back to the earliest years of the discipline and continue to have resonance. Others, such as spatial analysis and structuralism, had major impacts on reshaping the direction and methodologies of the discipline and, although now contested, retain a presence and a strong heritage.

Within this pattern of paradigm change, many of the so-called systematic (or adjectival) geographies have persisted and flourished. Human geography became organized around such systematic themes as historical geography, urban geography, economic geography, political geography, and population geography, both for purposes of delivering the undergraduate curriculum and in order to bring together academics with shared research interests. These divisions with their descriptors are often retained but have tended either (1) to subdivide further into more specialized areas, or (2) to form new groupings that work on the interstices and overlaps between the systematic geographies (Figure 12).

The most recent innovations, such as postmodernism and critical theory, have produced the diversity that now typifies the discipline. On the one hand, they have prompted powerful intellectual debate; on the other, they have shown little interest in either the traditions of human geography or different approaches.

The 'cultural turn'

A useful exemplar of the ebbs and flows in human geography since the 1980s is the so-called 'cultural turn'. It has become a major force for change, particularly in the United Kingdom and some other parts of the English-speaking world. The term 'cultural turn' has been used to describe a fundamental shift in approaches to the study of cultural geography (see box). This, however, has not been the limits of its influence as it has impacted on many branches of human geography, such as economic and political geography, and has subjected their objects of study to a greater consideration of cultural and historical specificity. The essence of the cultural turn, then, is that it suggests that large swathes of human geography must be recast in a similar mould. This is by no means a generally accepted position, and in many parts of the world, including the United States, the cultural turn has been much more muted

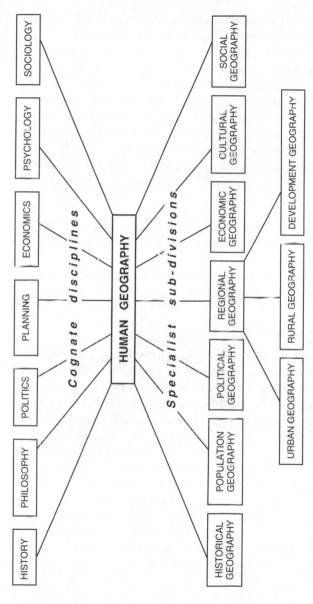

12. Human geography: its specialist subdivisions and links to cognate disciplines

Cultural geography

Traditional cultural geography was strongly embedded in the relationships between people, settlement, and natural environments; the term 'cultural ecology', which was often used, signified this interaction. Culture itself is a difficult term to define but is mainly summarized as way of life with both material (buildings and artefacts) and non-material (language, religion, and customs) expressions. Cultural geography developed concepts such as culture area and culture region to identify territories characterized by common cultural traits.

New cultural geography has moved away from expressions of culture to a much stronger focus on the meanings and values that underlie these artefacts and activities. Cultural geographers have been drawn much more into theories of language and studies of signifiers and symbols. All forms of representation, including art, architecture, fiction writing, film, and music, were afforded significance in the sense that they could be used to derive meanings and insights. Don Mitchell (the American geographer) concluded that the focus is on the ways particular social relations intersect with general processes, grounded in actual places and the social structures that give them meaning.

Non-representational theory is a new approach that stresses performance and embodied knowledge; it moves towards interpretations of culture based on practice. The focus is not upon the outcomes of cultural processes but on the performance and enactments that lead to those outcomes.

in its impact than in the United Kingdom. Nevertheless, some have spoken of 'culture wars': cultural geography is not a quiet land.

Turn from what?

Cultural geography has an established presence in geography. For much of its history it was founded on objectivity, the study of artefacts, and the visible imprint of people on landscape. Figure 13 depicts a landscape from rural Wales and from it can be 'read' the key features of its evolution for human settlement. It is a dispersed form of settlement with scattered farms in enclosed fields. It is mainly pastoral, as shown by the type of livestock evident in the fields. Some of the key service functions, such as the school and chapel in the foreground, are isolated from the people they serve but act as places of conflux. The high pastures, lightly covered with snow in this photograph, will be used as summer pastures to which stock are moved at that time of year. In the Welsh rural tradition, there was a *hendre,* or home farm, and a *hafod,* or summer dwelling only occupied at that time of year. All this and much more can be read into landscape.

From the early 20th century onwards, Carl Sauer, the leading cultural geographer of the day, and his 'Berkeley School' developed the whole spectrum of cultural geography and related cultural ecology. It offered evidence of the differences between cultural groups and their relationships with environments. Sauer's general thesis was founded on these tenets and detailed studies of landscape and its artefacts, such as the tobacco barns of the mid-West and the 'markers' of the Mormon culture region, such as churches, wide streets, and distinctive hay-ricks. Culture was often defined as a way of life and landscape as palimpsest was the mirror in which this way of life could be seen and its heritage traced over time. Sauer and his followers were less concerned with the 'inner workings of culture' than with the outcome of culture as it worked on the world.

13. A rural landscape in mid-Wales

Key features of the new cultural geography

The new cultural geography of the last part of the 20th century was precisely concerned with the inner workings of culture; the very component that traditional cultural geography had chosen to ignore. New cultural geographers wished to distance themselves from notions of culture as objective reality, visible and material. They believed in culture as a process that was socially constructed, actively maintained by people, and supple in its engagement with other spheres of human life and meaning. The new cultural geography was interested not in the patterns of artefacts or of forms of behaviour but in the meanings that underlay these objects and activities. It has been described as the medium through which people transform the mundane phenomena of a material world into a world of significant symbols to which they give meaning and attach value. All acknowledged the influence of Raymond Williams, the British social commentator, who described culture as the signifying system through which (though among other means) a social order is communicated, reproduced, experienced, and explored. Culture can therefore be seen as a set of signifying systems with texts capable of multiple readings. Thus, the subject matter of cultural geography changed from objects and things, such as tobacco barns and agricultural implements, to all forms of representation, including symbols, gestures, words, and artistic expression (such as art, fictional literature, and dance): that is to images that *create* meanings.

What lessons can be learned from this brief review of cultural geography? The new cultural geography has enriched the wider discipline of human geography. A defining characteristic has been to seek a fully dimensional cultural geography approach. However, the older cultural geography has not been replaced, and the two approaches proceed side by side, often co-existing in relative independence of each other. The social philosophies from which the new cultural geographers draw their intellectual stimuli have little relevance not only for the older approaches but also for physical geography as practised today.

Noel Castree, a British geographer, captures the sense of concerns about this change, which are widely held: 'In little over a decade, explanatory and moral solidities have melted into air, only to be replaced by a plethora of alternative philosophies, theories and techniques.' The claim that the new cultural geography extends throughout the whole of human geography as a modifying and redefining force can be overstated. It can also be argued that new cultural geographers fall into the fallacy of assuming that social constructions have more substance and causal influence than may be the case. Finally, the cultural turn has introduced new subject areas, categories, and lenses through which the world can be seen. Issues such as racism, feminism, and sexuality have gained new footholds in human geography and this has undoubtedly been one of the positive impacts.

Introducing new issues: redefining others

The emergence of issues such as racism, feminism, and sexuality as key areas for study in human geography can be linked to the focus on meanings and values that underlie places and activities that was so strongly promoted in the new cultural geography. Additionally, their emergence is a reminder that human geography was not working in a vacuum and that similar trends were being experienced across a range of disciplines, particularly in the social sciences. We use the word 'emergence' but in some ways this is misleading. Social geographers, for example, have an established research interest in ethnic segregation and the ways it changes over time. Similarly, there are studies of the elderly in the city that long pre-date new cultural approaches. Analyses of old people in an American city described them as 'prisoners of space' whose perceived life space was limited to home, to immediate neighbourhood, and to remembered places from the distant past.

Gender and sexuality

The feminist movement was particularly strong in the last quarter of the 20th century and among its relevant questions was that surrounding the roles of women in spaces and places designed primarily by men of a particular class and ethnic designation. Many parts of the central city retail area, for example, are widely used by women and yet they had little input into the design and planning of such areas. Open spaces, parks, and woodland are regarded by men as prime recreational and leisure areas, and yet for women, especially in the darker hours, they are often unsafe places to be avoided or used in fear. This notion of vulnerability can be extended to other groups such as the elderly, the disabled, and children.

Similarly, it is clear that heterosexual norms tend to be applied to the organization of space, and for gay people there are again difficulties of design and access that have to be overcome. It has been shown that simple exchanges of affection, such as embracing, by gay couples are not tolerated as public acts in many predominantly heterosexual societies. Public space imposes its own codes of acceptability and overtly gay behaviour can lead to victimization and abuse. One response is to create gay spaces, another may be to participate in activities such as parades that make a statement about sexuality. Other spaces, such as those used by the Women's National Basketball League in the USA, are often perceived as 'lesbian' spaces when in fact their primary role is as a focus for women's sport; the lesbian connection is at most partial.

In other words, the spaces and places so often treated in a uniform way by human geographers had different meanings for different types of people. These social worlds held a range of emotions and values that needed to be understood. Awareness of these kinds of considerations was one of the ways in which the tenets of the new cultural geography permeated across the spectrum of human geography.

Ethnicity and race

There is a long tradition of the study of ethnic segregation in human geography. It has mainly followed the path of developing techniques such as location quotients and indices of dissimilarity to measure the extent of ethnic residential segregation in cities, with a flurry of analyses after each national census. Such studies have always had their explanatory dimensions that sought to balance and understand the varying influences of constraints, on the one hand, and choice, on the other. Why some ethnic groups strive to remain segregated is an interesting question. The answer, for the New York Chinese communities at least, seems to lie in their wish to maintain the language and customs of their original areas of residence.

Segregation studies have been linked to processes of migration and to the growing concentrations of poverty within urban areas; poverty areas tend to be linked with recent migrants and to certain ethnic groups. The word 'process' is important in this field of study, as change is a feature, and the designation of the 'truly disadvantaged', or the residual groups unable to improve their quality of life and left behind in the least desirable neighbourhoods, captures something of this urban dynamic.

One outcome of the new focus on meanings is greater awareness that trends of segregation were being studied primarily from the viewpoint of the dominant white charter group. Today, another question is asked: What is the perception from within the ghetto or the emerging ethnic cluster? Terms such as 'social exclusion' have gained prominence and are applied across many minority groups. Thus, it can be argued that diverse ways of being in the world should be seen as legitimate and that 'out-groups' should not be stigmatized or excluded from the mainstream. There is an interest in the concept of diaspora, or the dispersion of specific groups, and the role of the host city as a site of memories, emotions, and sense of belonging.

In the traditional model, much emphasis was placed on assimilation as a process, and it was often argued, for example, that it took three generations to assimilate new immigrants to their host society. The model offers an example of the thrust of the empowered parts of society to reduce or eliminate differences and achieve conformity. It ignores the fact that full assimilation is not possible for poor immigrant and minority groups (though distinctions in the assimilation process between, for example, cultural and economic assimilation have long been recognized). Whereas one favoured approach might be to tolerate difference and diversity in a multicultural society, political circumstances argue against the idea of societies within societies, or subcultures, that might threaten the existing hegemony.

Changing systematic human geographies

What we have termed the systematic human geographies, such as economic geography, population geography, and historical geography, are still in common use but have often been replaced by much more focused theme-based approaches. A brief look at what is available in departments of geography in British universities shows that these themes are reflected in course modules currently available, such as urban services; global issues and problems; geography of money and consumption; memory, space, and place; cyberspace geographies; and the geography of affect and emotions. The plural term 'geographies' is often used to emphasize the variety of approaches that now exists.

Some commentators on the roles of more traditional systematic geographies argue that they have experienced the wider impact of the 'cultural turn'. The economy, for example, is no longer a self-evident object of study as economic and cultural practices increasingly 'bleed' into one another and intermix. A relational approach would integrate economic, social, cultural, institutional, and political aspects of human geography and move away from the neat systematic categories. This interpretation, however, is contested: many economic geographers ignore the debate surrounding the claims of the cultural turn. Furthermore, in Britain at least, economists have increasingly taken up the rigorous methods of spatial science and it is they who offer a 'new' economic geography that is evidence-based and tackles existing problems in society.

Geographies of development have a strong economic content but also demonstrate the importance of recognizing the power of cultural and political factors. Traditionally, development geography studied the disparities between different countries of the world and the causes that underlay these disparities. There has long been a distinction between More Developed Countries (MDCs) and Less Developed Countries (LDCs), with the addition of Newly Industrializing Countries (NICs) to recognize change. Countries were placed in categories defined by measures such as Gross Domestic Product (GDP) and Human Development Indices that combine measures such as life expectancy and literacy levels. Causes are many, but there are wider explanations such as dependency theory, which suggests that LDCs continue to be exploited by MDCs and global corporate investment. Organizations such as the United Nations and the World Bank are major players in attempts to reduce disparities, and there are detailed studies of debts carried by LDCs. The debt burden remains and is exacerbated by civil wars, famines, and the onset of pandemic diseases such as AIDS. The politics of development is also a major issue that covers both the 'benevolence' of the MDCs and the attitudes of LDCs towards change. Questions of sustainability figure increasingly in development geography as the tensions between development and the needs of the environment become clearer. Here again, MDCs and LDCs may have quite separate agendas. There is also strong awareness of the need to avoid the abandonment of local modes of agriculture and native economies and to use the consultative channels that are available.

Political geography has moved from geopolitics and the idea of centres of global power to studies of identities, empowerment, resistance, mobilities, and differences, and the ways in which these are played out in geographical space.

Urban geography has in some ways mirrored trends elsewhere and has been strongly influenced by, for example, humanistic geography that brought in studies of themes such as graffiti as a territorial marker and cultural geography with its interest in the affective values and emotions attached to city neighbourhoods. Broader trends such as the growth of gender and sexuality issues in geography have prompted studies of the ways in which different groups occupy urban space. On the conceptual front, the studies of cities and urban life have attracted the application of postmodern ideas, post-structural theory, and critical discourse analysis.

How do we move from the conflicting positions that often emerge to some kind of convergence? A perceptive and compromising comment from a British geographer is helpful:

> Although this book is titled Social Geographies it makes no claims to occupy a discrete intellectual space, which can be identified or sealed off from other traditional sub-disciplinary areas such as cultural geography or political geography. Rather the plural social geographies, which emerge here, are a porous product – an expression of the many connections and inter-relationships that exist between different fields of geographical inquiry.
>
> G. Valentine, *Social Geographies: Space and Society* (2001)

The sentiment is one of 'openings'; a desire to define and defend a particular style of approach and theme, but also a willingness to recognize and learn from other approaches. It also serves as a statement on the current 'state of play' and offers a good basis from which human geography can progress.

Whither empiricism and evidence-based studies?

Aspects of a general debate

A distinguishing feature of human geography has always been its strong empirical content, which is a feature of both its research and its applications in the real world. The oldest traditions such as exploration, discovery, and fieldwork all involved the careful collection of information and data. Much of this was qualitative and descriptive and formed the bases of expedition reports and regional studies; much was quantitative and measurable such as the data inputs to cartography, charts, and mapmaking. Spatial analysis and the 'quantitative revolution' of the 1960s brought all of this to the fore and prompted human geographers to make use of the many different data sources then available. As satellite imagery and geographical information systems have developed in their applications to human geography, the range of data inputs and their analysis has once again increased.

Against this continuing empirical tradition in human geography and its expressions in evidence-based research are two major tensions within the discipline that work in contradictory directions. First, the concept of data and statistical analysis is anathema to many adherents of the new cultural geography. The 'quantitative revolution', or shift to a more scientific and measurement approach, is regarded by them as the 'dark ages', a 'diabolical science', and a low point in the emergence of human geography. Their approach, founded on abstract ideas and the qualitative search for meanings, often has little time for data and certainly not for measurement. Second, the grand theories or meta-narratives found in the various forms of structuralism similarly had little time for empiricism and case studies. During the 1990s, when qualitative case studies of the real world became the dominant approach, there was a strong reaction against what Marxist geographers labelled as the 'empirical turn'.

Both of these positions need qualification. Cultural geographers may disown the value of statistics or social surveys but would claim that their quest for the meanings behind areas, landscapes, and activities illuminates our understanding of the world. The grand theorists would argue that their interpretations of hidden structures allow causality and explanation to be achieved. Yet there are significant costs in moving away from the empirical traditions of human geography. One is relevance and another is involvement. As the Canadian geographer Derek Gregory put it:

> If we do not care about the world, if we treat it merely as a screen
> on which to display our command of high technique or as a
> catalogue that serves to furnish selected examples of our high
> theory, then we abandon any prospect of a genuinely human
> geography.
>
> D. Gregory, 'Geographies, Publics and Politics' (2005)

Continuity in evidence-based research

Thankfully, the empirical tradition and evidence-based research that has application to the problems and issues facing society is far from absent in modern human geography. The detailed studies of census data by geographers continue to inform us on the nature and extent of population change. Both the demographic structures and population distributions in most countries are far from static and an understanding of these facts is essential, for example, in the provision of public services. Migration and movement have always typified human populations and very significant population movements continue to affect many countries. New confederations, notably the enlarged European Union, have enabled large-scale labour migrations that often lead to permanent residence. Retirement migration in Europe and North America has become a major feature. Displacement of people and the turmoil and friction that seem endemic in many societies have produced the relatively new phenomenon, in scale at least, of asylum-seekers searching for a better and safer life in more advanced societies.

Human geographers have studied and interpreted both the processes of change and their outcomes. Urbanization and the emergence of large cities continue to be a strong theme within human geography. Studies of *globalization* have prompted major debates at both a conceptual and an empirical level. The Canadian geographer Wayne Davies defined globalization as:

> The increasing global spatial flows, interdependence of people, information, goods, organisations and states that are connecting people and places at a world scale, and are creating changes in the structures and organisations of society and places.
>
> W. Davies, 'Globalization: A Spatial Perspective' (2004)

One of the most visible impacts of globalization is the spread of the transnational companies, with brands such as Coca Cola and McDonalds achieving a ubiquitous appearance in cities across the world. There are key issues raised by globalization. Do these ongoing global forces for uniformity submerge local and regional differences? Are local histories and cultures sufficiently robust to retain clear identities? Is it the interests of the main political and economic power brokers that are driving the globalization agenda forward? Is it selective in its promotional and control strategies? As the British geographer Doreen Massey has pointed out, the concept of free trade in a world without boundaries is contradicted in areas where strict controls continue to exist. There is danger of what she terms a 'duplicitous manipulation of geographical imaginations'.

Urban geography has a long-standing interest in the uneven distributions of advantage and disadvantage within cities.

There have been studies, for example, of poverty areas, areas of multiple deprivation, and areas where the incidence of crime appears to be disproportionately high. Standard approaches have included the search for territorial indicators that allow some measurement of the extent of concentrations and the range of problems that are located in the most disadvantaged neighbourhoods. Census small area statistics have provided the main data sources and the fine-tuning of these data has moved from wards to enumeration districts and to postcode areas. The ongoing debate about the 'postcode lottery' refers to the human geographical fact that your life chances often relate to where you live; some neighbourhoods are privileged whilst others are deprived. Beyond the measurement of concentration and the classification of neighbourhoods into a typology based on selected indicators, geographers have moved to more in-depth analyses of the quality of life within specific areas. Generally the issues relate to the specificity of place and its influence; whether there are subcultures or local sets of values and behaviour within the more general traits of society.

Allied to more traditional studies of poverty areas have been analyses of financial exclusion and the ways in which sections of the population are excluded from access to loans, mortgage funds, and credit. Again, there are geographical studies of the meanings of place and the existence of affective values attached to specific neighbourhoods that may protect them from the thrust of market forces, at least in the short term. The classic study of Boston's Beacon Hill, where for many years resident groups were able to resist the incursion of commercial activities and maintain the character of their neighbourhood, was an early example of this approach; as are more recent similar studies of Shaughnessy Heights in Vancouver. At another front, here is a new focus on consumption that moves beyond established approaches to retail centres, shopping patterns, and service provision to less well understood worlds of second-hand markets, informal sectors, and car-boot sales.

14. **Gentrification of the inner city: Elder Street, Spitalfields, London, UK, where formerly run-down houses have been upgraded and improved for the professional classes**

Gentrification is a process whereby older residential areas, especially in the inner city, are refurbished and moved upwards in the cycle of housing change and regeneration. Figure 14 shows a gentrified street in the Spitalfields district of London. The three-storey houses have been re-furbished and modernized within the original facades. Much gentrification takes place in the private sector, though there are major public-sector initiatives, often linked with flagship projects such as economic regeneration of the inner city, major sports or cultural facilities, and docklands projects. The process has encouraged a lively debate about causes that covers many aspects of the interests of human geographers. Is gentrification a product of changing choices in the residential market place and thus driven by consumer preferences? This is partly true, and a changing labour market with its focus on service jobs and a greater representation of women in the professional workforce is one driver in the process of change. The downside is that those able to make choices are the prime movers and there has been a significant displacement of low-income households with the upgrading and re-costing that occurs under these schemes. The other key factor is structural change and the investment strategies of large agencies motivated by profit returns. Gentrification is a complex process and involves residents, past and future, landlords, investors, financial institutions, planners, and municipal authorities. It is the latter who have to mediate between the ambitions of developers and the needs of local populations.

Regional geography has been a traditional part of human geography though it belongs to geography as a whole. Regional geographies tend to reflect analysis at a meso-scale with a concern for aggregates and often descriptive narratives. Cultural geographers have attempted to move the scale to the micro, with an emphasis on body and self and the primacy of the individual. It has been argued, for example, that traditional concerns for regions, places, and landscapes became mediated through those for self and body. It has also been proposed, for example, that

Jean Baudrillard's use of the imagery of signs and symbols in the American landscape, such as billboards and some types of architecture, is now a more valid representation of a 'region'. A qualitative and impressionistic approach that allows meanings to be read into the 'text' is preferred to one that involves empirical description, data, and measurement. As a counterpoint, Christopher Butler, the Oxford historian, describes Baudrillard's assertion of the essential unreality of the culture in which we live as outrageous. Clearly, there is room for debate! The progress of the region can be traced from the notion of differentiation of territory, to environment and *genre de vie*, to the functional space of regional science and the meanings of place and social relations in social and cultural approaches. Regional geography vies with the issue of the 'local' in a globalizing world, and this is one of the many areas in human geography where the evolution of thinking has provided radical change.

Chapter 4
Geography as a whole: the common ground

The picture that emerges from our discussions so far is of a diverse field of study. The description 'a broad church' is often applied and is very apt for the range of topics and approaches currently investigated by physical and human geographers. It is the purpose of this chapter to search for and demonstrate the common ground, the bonds that hold geography together as a single discipline and provide it with a unified identity. We first briefly set out the most important shared concepts and practices, and then focus on five areas of research, scholarship, and study that we term 'integrated geography'. This demonstrates by example the distinctive and important role played by geography as a whole.

Shared concepts and practices

The unifying aspects that hold geography together can be set out as follows. First, there are the core concepts that we have identified as space, place, and environment. Allied to these is a set of generic concepts that serve as common currency within geography, including time, process, and scale. It is important to recognize that geography has no monopoly over these concepts (even those described as core). They are used across wide fields of knowledge, but it is the manner of their use by geographers that gives distinctiveness.

It was the founding practices of geography – such as regionalism, historical geography, and environmentalism – that led to the recognition of geography as a subject worthy of study. These are underpinned by geography's focus on both nature and culture, and hence its ability to act as a bridge between the sciences and the humanities. Today there are many key research questions that can only be addressed by combining physical and human geography: examples of these, such as the questions posed to society by the exploitation of resources, natural hazards, and global environmental change, are detailed in the second part of this chapter. There are, moreover, parts of geography, perhaps best exemplified by the idea of landscape, which require a unified approach. This too, is developed in a separate section below. Finally, continuity of geographical research and education in universities and schools depends upon the maintenance of geography's single-subject identity.

Integrated geography

The term 'integrated geography' recognizes explicitly those aspects of the discipline that include both the physical and human dimensions. There is a certain irony in the fact that at a time in its history when there is ever more specialization, the need for integrated geography is greater than ever. This requires renewed recognition of the distinctive qualities and intellectual heritage of geography as a whole. The aim here is neither to mount a defence of some past 'Golden Age' of geography when integration was the norm, nor is it to ignore the value of modern developments. Rather, it is to demonstrate that there are many integrated approaches to the study of geography that have proved capable of being adapted to modern issues. In each of the five fields of integrated geography to be discussed, interaction between physical and human geographers has persisted. Various combinations of geography's core and generic concepts are emphasized; and the impossibility of answering certain important research questions about the Earth and its peoples without an integrated geographical approach is illustrated.

Regional geography

Regional geography can be defined as place description, analysis, and synthesis. It dominated geography in the early to mid-20th century and as such is often viewed as a phase in the development of the discipline. Regional geography is not, however, merely of historical interest as it continues to play a central role in relation to research and policy. Geographers remain committed to regional research; to analysing and explaining regional differences; to testing general theories in the regional context; to developing policies for particular regions; and to solving problems in specific places. Indeed, no fewer than nine Speciality Groups of the Association of American Geographers are essentially regional in focus, covering America, Canada, Latin America, Africa, Europe, Asia, China, and Russia (with Central Eurasia and Eastern Europe). In addition, most geographers, even those who would not claim a regional focus to their work, in practice carry out research on limited areas of the Earth's surface about which they are best informed.

How does modern regional geography differ from traditional regional geography? It can no longer be characterized by the 'areal differentiation' of the American geographer, Richard Hartshorne, or the regional descriptions of the British Naval Intelligence Handbooks, or indeed of the later textbooks that they inspired. The implication of all these works was that each region was unique and relatively homogeneous with closed boundaries: the distinction was not always clear between the regional reality, where one region might merge imperceptibly into another, and the regional method, which could be applied anywhere. Modern geographical studies of regions are not studied in isolation but take account of multi-scale relationships with connections up to global scale. Regions are now delimited for particular purposes according to specific criteria: they tend to be regarded more as methodological devices, rather than generally distinguishable parts of the Earth's surface.

What kind of research is involved in modern regional geography? This complements the emphasis on generalization and globalization with a consideration of the specificity of place, the local effects of global processes, and locally generated processes. Distance is no longer important for some global social and economic activities but, at the same time, local forces are perpetuated and new ones generated by cultural differences. The particular problems on which regional geography focuses are often associated with spatial inequalities and uneven development with important implications for political agendas and government policy.

Geographical studies of Africa illustrate some of the contemporary research themes of regional geography in the context of the developing world. The simple theory that human population growth leads to environmental degradation and agricultural decline, for example, has been shown to be an oversimplification. Many factors of the biophysical and human environment (the latter including, for example, market access, land tenure, technological innovation, and politics), interact in complex ways to determine whether a production system is sustainable. Where these environmental conditions are favourable, imported production systems based on advanced technologies and financed by massive injections of foreign aid to governments may produce rapid and sustainable agricultural intensification. Such schemes often fail, however, due to unfavourable conditions. In these different circumstances, indigenous systems and/or the provision of micro credit facilities directly to the poor may provide a more appropriate basis for development. State intervention in land reform, multinational corporations, and access to local resources, indigenous rights and wildlife conservation, agrarian and pastoral conflicts, and changing women's roles in rural and urban life are

15. One of 380 families of herders sheltering in the El Hache Camp, North Eastern Province, Kenya, after losing their livestock to the severe drought that extended across East Africa after the failure of the rainy season in October 2005, continuing a decade of low rainfall

other geographical issues involving resources, environment, and development.

Hunger and famine in sub-Saharan Africa is a regional problem where particularities of the natural and social environment are increasingly being seen in the light of the values and perceptions of a new world order (Figure 15). Understanding of the particular physical basis of the problem is important, including the rainfall regime, the changing frequency of droughts, soil quality, and the habitat conditions under which disease vectors thrive. But so are increasing population pressure and its ecological impact, effects of the slave trade and the colonial era, divisive internal ethnic and religious differences, globalization and outside political interference, and the widespread failure of the African political leadership. There are similar geographical dimensions to the very different problems that are posed in other regions of the world, whether these involve the deforestation of Amazonia, the future of Antarctica as an international reserve, the rise of China as an economic power, the spread of terrorism from the Middle East, or the conflicting forces of centralization and devolution within the European Union. All the characteristics of geography are necessary to understand the complexities of regional problems.

Historical geography

Essentially, historical geography is the geography of the past. As a field of integrated geography, the key geographical concepts of space, place, and environment are considered in the context of past times. This often involves analysing a particular place or region at some time or period in the past (a 'time slice'). A classic example of this is the painstaking research on the Domesday Book carried out by H. C. Darby based in the Department of Geography at Cambridge University. The Domesday Book contains a mine of information on the geography of England in AD 1086, shortly after the Norman conquest. From this source, Darby quantified and mapped the human population, the area and uses of woodlands, the numbers of farm animals, and regional variations in the economy, including incomes.

A second approach involves using evidence from the past to help understand the present-day world, including the recognition of those phenomena of the present landscape ('relicts') that have been inherited from the past. Examples of such relicts in the present landscape include 'U'-shaped valleys inherited from glacial times; transport routes that follow the course of Roman roads (such as the Fosse Way from Exeter to Lincoln); the Norfolk Broads, once thought to have been natural lakes but now known to have originated through peat-cutting in the 13th and 14th centuries; and the many features of urban and industrial areas that form the modern heritage of the Industrial Revolution.

A third approach investigates changes through time in a particular phenomenon or a whole landscape (a 'time sequence'). Here we may use the example of the history of the Norse settlements in Greenland, which was colonized from Iceland around AD 985 and lasted for some 500 years. The Western Settlement lasted until the mid-14th century whereas the people of the Eastern Settlement died out towards the end of the 15th century. The decline coincided with the 'Little Ice Age', and it is an intriguing question as to whether there was a causal relationship. Climatic conditions certainly deteriorated: the graves of those who died were subsequently entombed in permafrost that only thawed in the 20th century. Crop failures would have increased in frequency as climate deteriorated, and connections to the outside world by ship were made more difficult by the extension of sea ice. However, the precise causes have yet to be established. Other factors that have been implicated include: overpopulation, soil degradation, and erosion leading to decreasing yields; the inability of the settlers to change their cultural values and lifestyle and thereby adapt to changing conditions; conflict with the native Inuit population whose economy was based on the marine ecosystem and was more sustainable; congenital infertility in an in-breeding population; and declining trade with Europe.

Like regional geography, historical geography has diversified considerably from merely providing a descriptive catalogue of historical change. Close relationships with history remain and there is a sense in which all geography is historical geography. Most human activities and natural phenomena that occur on the Earth's surface are of potential interest to modern historical geographers. In focusing on natural environmental change and the evolution of the natural landscape, physical geographers who investigate the historical dimension have largely done this without reference to historical geography. The recent tendency has been for cultural geographers to dominate historical geography with an emphasis on the different ways that places and landscapes can be shaped and experienced by people differing in social background, nationality, ethnicity, class, income, gender, or age.

The shift in modern times, therefore, has been from a focus on materiality, physical forms and artefacts, to one based on cultural processes. There are new emphases on the symbolism of landscape, for example, and the meanings with which it is imbued. Historical geography has always had its own diversity, and this is maintained in recent projects such as studies of the plague of AD 1665, negotiating colonialism, the architectural history of London Bridge, and Pierre Nora's sites of memory. However, the British historical geographer Michael Williams, in a review of the evolution of historical landscapes, stresses the roles of class relations and of alternative 'ways of seeing' the landscape. His plea is for a reconciliation between contrasting approaches by accepting that landscapes comprise both tangible and intangible nuances and realities. As with other forms of human geography, this is another example of the need for more traditional approaches to vie with the new focus on symbolism, meanings, and values.

The approach of understanding human–environmental interaction through time provides a particularly pertinent illustration of integrated geography. This presupposes that both the biophysical and the cultural changes are known or can be reconstructed over the same timescale. There is a long tradition of reconstructing the biophysical environments associated with successive phases of human occupancy of different parts of the Earth's surface and of inferring how people impacted on their environment during forest clearance, settlement, land drainage, rural transformation, urbanization, industrialization, and trade. Reconstructing the former natural environments of prehistoric and later peoples often involves scientific expertise from physical geography, environmental archaeology, anthropology, and other sciences. This is nevertheless a relatively easy task compared to the attribution of causes and, especially, understanding the human decision-making processes involved in past times. Human impact was variable in space and time, and was conditioned by, amongst other factors, people's changing perceptions of the natural environment and technological change. Furthermore, conceptions of nature and ideas about relationships of societies to natural environments have themselves changed and have a history as well as a geography.

Geography of human–environment interaction

Reciprocal interaction between the natural environment and people is the key concept of an integrated geography. It provided the strongest rationale for establishing geography as a separate discipline during the 'Geographical Experiment', and continues to provide a strong academic justification for physical and human geography remaining together in the same university department. Two major interconnected sub-themes can be identified: first, studying the complex effects of different natural environments on societies and their activities; and second, understanding the nature and extent of beneficial as well as adverse human impacts in different environments. Both sub-themes presuppose sufficient knowledge of the relevant physical and human geographical patterns and processes operating in both the natural environment and human society.

Attempts by geographers to conceptualize and theorize how environment interacts with society have had mixed success. The historical role of environmental determinism in the early 20th century, which portrayed environmental effects as simple, direct, causal links between, for example, climate and human characteristics or the decline of civilizations (Figure 16(A)), was fundamentally flawed. Since then, the complex, indirect, and reflexive nature of the environmental relations of society has been recognized and alternative, more sophisticated models have been proposed, two of which are shown in Figure 16 – (B) and (C).

A Environmental Determinism

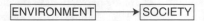

B Technological Materialism

C Adaptive Systems

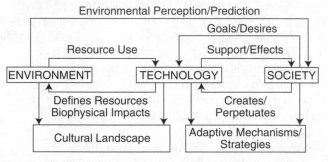

Geography as a whole: the common ground

16. Three models of environment–human interaction:
(A) 'environmental determinism'; (B) 'technological materialism';
and (C) 'adaptive systems'

'Technological materialism' recognizes the important role of
technology in mediating both the way the environment affects
society and how society affects the environment. The way people
view their environment often reflects the extent to which the
exploitation of resources is possible given the technological
aids available to them. In other words, technology can be an
enabling factor. Invention of the plough, for example, enabled
agrarian societies to intensify land-use and increase productivity.
This in turn led to greater human impacts on soil fertility and

erosion. Similarly, almost every technological innovation has the potential to affect human–environment relationships in some way.

In the 'adaptive systems' model interactions between environment and society are mediated by many more social, cultural, economic, and political factors. Thus there are many reciprocal relationships with feed-back and feed-forward loops, as indicated by the direction of the arrows between compartments in Figure 16(C). These features of the model reflect the complexities of the ways environments are perceived and used, and the ability of society to develop adaptive mechanisms and modify their strategies over time. Human adjustment to flood hazard provides a good example. In modern Western societies this, more often than not, takes the form of engineering schemes that protect against floods likely to recur once in a century or two. This solution has been adopted as a result of the interaction of social demands, economic costs, and political pressures. It represents one possible response to flooding – not necessarily the optimum strategy – that has evolved over time.

Geographical work focusing on the effects of environment on society contributes, for example, to understanding the exploitation of natural resources, and the vulnerability of people to natural hazards. A natural resource includes anything in the natural environment that is capable of exploitation by society, but what is exploited as a resource in a particular place depends not only on its availability but also on what that society values and chooses to exploit. Societies in different places or at different times may perceive resources differently because of different cultural values, levels of technology, or economic or political considerations. The position of wildlife is a good example in this respect: to some it is regarded as a source of food, such as 'bush meat', whereas to others it must be preserved for posterity or exploited in different ways by tourists. The geography of natural resources therefore draws on both the biophysical nature of the resource and many aspects of the associated human environment.

The distinction between renewable and non-renewable resources is important in this context. The former, such as soils, fresh water, forests, and fisheries, are regenerated by biological or environmental processes and may be harvested indefinitely provided that the sustainable yield is not exceeded. But the exploitation of such resources is increasing at a faster rate than the world's population: since the 1950s, world demand for water has tripled, catches of fishes have quadrupled, and consumption of food has increased six-fold. Technological innovations in agriculture, water supply, forestry, and fisheries have proved capable of both increasing yields and exceeding sustainable yields, which can lead to resource depletion and the degradation of whole geo-ecosystems. There are often knock-on effects throughout economic and political systems, as exemplified by so-called 'Water Wars' in the Middle East and elsewhere, where upstream abstraction of water from rivers and groundwater has led to limited downstream supply. Reserves of non-renewable resources, such as fossil fuels and metal ores, which are in limited supply because of their slow rate of formation by geological processes, may be depleted and exhausted (although recycling is possible in some cases, notably metals). Technological change and/or changing societal values may lead to an increase in exploitation but they are also capable of increasing reserves, reducing usage or creating substitutes, which reduce the rate of depletion and decrease the likelihood of exhaustion. The exploitation of renewable and non-renewable resources therefore raises issues of production and consumption, management and sustainability, conservation and preservation, all of which have important geographical dimensions.

Several of these resource issues are exemplified by the use of groundwater in arid lands. On the one hand, technology has created a productive landscape and has caused 'deserts to bloom'. This is shown in Figure 17, where the circular productive areas produced by sprinklers from rotating pipelines contrast dramatically with the surrounding barren landscape. The downside of this admirable achievement is often the long-term depletion of the groundwater reserves at a faster rate than natural recharge of the aquifers from rain falling on the surrounding regions. This may require water rationing. Furthermore, salinization often results in reduced productivity and eventual desertification caused by the accumulation of salts following the evaporation of water from the soil. The United Nations Food and Agriculture Organization has estimated that 125,000 hectares of land are lost worldwide each year through salinization.

The vulnerability of people to natural hazards provides a second illustration of effects of the environment on people. Natural hazards are extreme natural events that pose a risk to human systems. They include meteorological, geological, and biological events, but human-caused pollution hazards and diseases that threaten human health are normally excluded. Severe impacts of geophysical hazards – such as earthquakes, volcanic eruptions, floods, and tropical cyclones – commonly have disastrous consequences for society. But the same level of hazardousness can have widely differing impacts in different societies with different vulnerabilities. In an average year, some 250,000 people die from natural hazards, more than 80% of these in developing countries. Recent examples were provided by the Boxing Day tsunami of 2004, which affected populations around the shores of the Indian Ocean, and the Kashmir earthquake in Pakistan in 2005. In contrast, economic costs in terms of damage to property and interruption to businesses tend to be highest in the developed world, belying the false notions that technologically advanced societies are the least vulnerable and that they are less vulnerable today than in the past. This was graphically demonstrated both by the inundation of New Orleans following Hurricane Katrina in 2005 and by the river floods in southern England caused by

17. Irrigation of the desert using pumped groundwater: (A) circular fields of green on the yellow sands of Saudi Arabia seen from the space shuttle Columbia; (B) a single field near Kibbutz Gvulot, North Negev Desert, Israel, showing the centre-pivot spray-irrigation system

prolonged rainstorms in the summer of 2007. Human populations and wealth continue to grow in hazardous places, driven by the need to exploit the resources available and often disregarding or being unaware of the risk. Integrated geography has the task of taking both the biophysical hazard and the cultural context fully into account.

Geography of global change

Spatial variation up to global scale, and temporal variability on timescales of relevance to human occupancy of the Earth, have long been major concepts within integrated geography. Thus, geography's interest in global change is not a new one. This interest can be traced within both of the previously discussed themes of historical geography and the geography of human-environment interaction. However, global change has recently become a dominant theme in its own right, largely because of broader concerns about the magnitude, rate, and direction of current changes in both the biophysical and human environment. On the one hand, global-scale human impacts on the biophysical environment now dominate to the extent that many believe this threatens the future existence of humanity itself. On the other hand, the globalization of communications, organizations, information, and other forms of human interaction have profound implications for the nature of economic, social, and political patterns within the human environment.

Global change has come to refer to the immediate past, present, and imminent future changes affecting the anthroposphere – the human-modified Earth's surface. The unprecedented rates of change during the Anthropocene – the last c. 200 years (see box in Chapter 2) – especially during the last 50 years, are illustrated by some key indicators of the natural and human environment in Figure 18. These global changes are driven, directly or

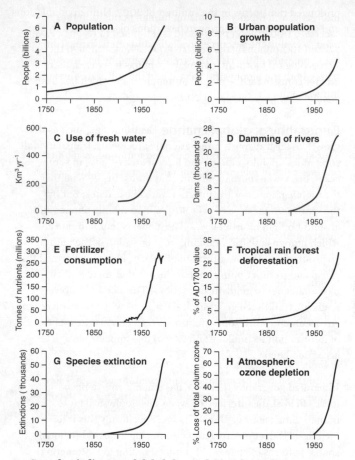

18. Some key indicators of global change during the Anthropocene: (A) world population growth; (B) urban population growth; (C) use of freshwater; (D) damming of rivers; (E) fertilizer consumption; (F) tropical rainforest deforestation; (G) species extinction; and (H) atmospheric ozone depletion

indirectly, by a rapidly urbanizing human population that now exceeds 6,000 million people and is estimated to reach between 7.3 and 10.7 billion by the year AD 2050. Thus, although the 'geo-ecological footprint' of each person has increased, there is greater concern for the overall human impact, which has been estimated to have increased 200% since 1960.

Rates of changes associated with the electronic age, such as the growth in air transport, international agreements, transnational companies, mobile phones, and internet users have, in some cases, been even more rapid and ubiquitous. In 1985, neither the mobile phone nor the internet were in existence; but by the year 2000 there were over 800 million mobile phones and around 1 billion internet users. The global extent and rate at which human activities, ranging from pollution of the natural environment to the organization of society, have transformed the world and its functioning systems means that there are few, if any, past analogues to guide future actions. This does not, however, eradicate traditional geographical concerns with space, place, and environment; rather, our attention is refocused as the realities of information technology and its variable impacts have to be accommodated.

Integrated geography therefore has a major role to play in this field at the interface between the biophysical and human dimensions of global change. Specifically this role includes: (1) documenting and monitoring local and regional spatial patterns of change; (2) understanding the interacting processes and explaining their effects in different places; (3) developing policies for the mitigation of environmental impacts at local to global scale; and (4) contributing to ethical frameworks.

A good example is provided by so-called 'hotspots' of biodiversity (Figure 19). They are defined as areas with exceptional concentrations of endemic species that are also threatened

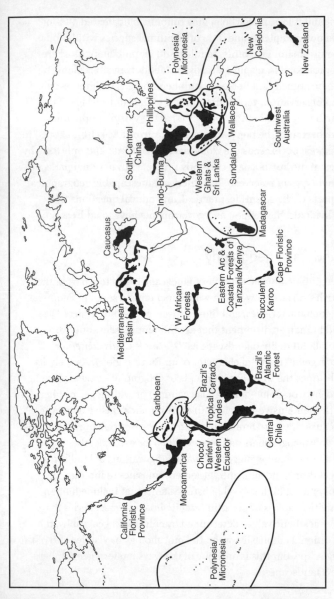

19. Biodiversity 'hotspots': the 25 leading hotspots, where large numbers of endemic species are under threat, are named and shaded

by habitat loss. As a whole, they contain an estimated 44% of all flowering plants and 35% of all animal species within four groups (mammals, birds, reptiles, and amphibians). Most are tropical forests, tropical islands, or Mediterranean regions. It can be argued that hotspots are where there is the greatest need for conservation, yet on around 12% of the land surface of the Earth they are home to about 20% of the world's population. Furthermore, the population growth rate of 1.8% (1995–2000) in the hotspots exceeds the global average of 1.3%; and only 38% by area of the hotspots are currently protected in national parks or other types of reserves. Thus, there are inescapable geographical aspects to the scientific, practical, and ethical questions that individuals, businesses, and governments have to address.

Landscape geography

Geographically, the concept of landscape refers to a part of the Earth's surface viewed as a whole, including a set of phenomena, their characteristics, and those aspects of the biophysical and human environment that are influential. Alexander von Humboldt defined landscape as '*Der totale Character einer Erdgegend*' (the total character of an Earth region). As such, it subsumes the three core concepts of geography – space, place, and environment – and can lay claim to providing geographers with their elusive 'object of study'. It remains elusive, however, because there are numerous different ways of viewing landscapes both by geographers and others. Thus, for example, they may be all of the following: particular configurations of landforms, vegetation, land-use, and settlement; mosaics of interacting ecosystems; higher-level holistic systems that include human activities; arrays of pixels in satellite images; or sceneries that have aesthetic values determined by culture. In geography, the traditional emphasis was on the morphology or visible form of the landscape, but this is only one of the ways modern geographers view landscapes.

Towards the middle of the 20th century, the German geographer Karl Troll was a major force in developing a geo-ecological approach to integrated geography based on landscape viewed as the product of natural and human processes. This can be seen as the initial stimulus to the late 20th-century emergence of the interdisciplinary field of landscape ecology, to which Russian, American, and Dutch geographers have also been major contributors. Landscape ecology has been defined as:

> ... the study of spatial variation in landscapes at a variety of scales. It includes the biophysical and societal causes and consequences of landscape heterogeneity.
>
> International Association of Landscape Ecology,
> *Mission Statement* (1998)

Ideas from landscape ecology have had a major influence on physical geographers as they have moved away from an emphasis on topography towards a deeper understanding of process interactions and how landscapes function and change as holistic systems. In geomorphology, for example, landscape change is seen as involving sediment budgets with various inputs and outputs of mass and energy; interactions between a range of Earth-surface processes, the substrate and the vegetation and soil cover; and erosional and depositional events that vary in magnitude and frequency through time. A good term for this is 'landscape dynamics'. Similarly, in their contribution to Quaternary science, physical geographers interpret their sections through sedimentary deposits and their sediment cores as a result of many interacting landscape processes forced by both natural and anthropogenic factors. Modern landscape ecology has been greatly affected and facilitated by the development of remote sensing and GIS. These technologies were designed for describing and analysing patterns and change in landscapes, and are highly applicable to landscape management and landscape planning.

Human geographers, for their part, are no longer concerned only with those aspects of landscapes reflected in material culture, but are now more interested in the underlying social, cultural, and political processes that produce landscape and with the meanings and values attached to it.

Human geographers still acknowledge their debt to Carl Sauer and his cultural ecology approach to landscape but now adopt a variety of interpretations. Landscape as palimpsest, for example, encourages the evolutionary interpretation of landscape. Landscape as taste and value focuses upon the alteration of landscape to reflect current vogues. Humanistic geographers seek to view interpretations of landscape by painters and writers as their ways of seeing. Descriptions of landscape as social process, as text, or as identity, reflect attempts to read into landscape the human forces that formed it. Landscape has become a concept that remains central to human geography. This is evident in the American geographer Denis Cosgrove's description of the concept as providing a focus on visible parts of the world, of suggesting unity and order in environment, and as a record of human interventions. Although some of the geographical interpretations of landscape can accommodate this diversity, the key challenge of integrating the physical and human qualities of landscape remains.

One of the most promising approaches for landscape geography is to build on the foundations of landscape ecology by investigating the complexities of landscapes as coupled natural and human systems. The integrated study of such systems can reveal patterns, processes, and surprises not evident when they are studied by physical or human geographers separately. This is exemplified well by interdisciplinary investigations in the Wolong Nature Reserve for endangered giant pandas in China. These studies measured variables that link the natural and human systems, such as fuelwood collection, as well as the more obvious aspects of the physical and cultural landscape, such as panda numbers and wildlife habitat, on the one hand, and human numbers and conservation strategies, on the other.

The giant panda is highly dependent on bamboo forest, which forms its habitat and supplies the bamboo leaves that are its staple food. As forests near the local households in Wolong were depleted by the collection of fuelwood for cooking and heating, fuelwood was increasingly collected from the bamboo forest, leading to the deterioration of panda habitat and threatening the panda with extinction. This led the Chinese government to establish the reserve and take other conservation measures to benefit both pandas and humans. Despite a reduction in the local resident human population due to their out-migration to work in cities, the demand for fuelwood increased, and panda habitat degraded faster after the reserve was established than before. This was partly the result of the consumption of local products by the large influx of tourists arriving from around the world to see the pandas. It was also affected by an unexpected increase in the number of households, each of which received a substantial increase in income from subsidies received as part of the conservation strategy. Household proliferation more than counteracted the reduced number of people per household, leading to an increase in demand for fuelwood, which further threatened the panda.

This example demonstrates that one type of integrated landscape geography can yield results that are both interesting and useful. It relates the spatial structure and underlying processes of the natural and cultural landscape in a unified way to the record of human interventions and attempts to reshape the world. It also goes some way towards capturing the essence of landscape and the essence of geography.

Shared future or separate paths?

Notwithstanding the mutual dependencies that have been emphasized in this chapter, it is clear from the recent history and current practice of physical and human geography that the discipline as a whole faces a dilemma. On the one hand, the shared past of physical and human geography, and the concepts they have in common, bind them together. On the other hand, the differences between them, in terms of subject matter and approach, which were highlighted in Chapters 2 and 3, respectively, suggest divisions and at least the beginnings of separate pathways of development.

The commonalities are most apparent between physical geography and those aspects of human geography that can be described as belonging to the social science tradition. The aspects of human geography that fall within the humanities tradition are more difficult to accommodate in a single intellectual framework with the natural science tradition of physical geography. This does not detract, however, from the fact that physical geography still has an essential role at the core of integrated geography. The dilemma, then, which is probably the main challenge facing geographers today, is whether and how the discipline of geography as a whole can hold together? This theme will be returned to again in the last two chapters of the book.

Chapter 5
How geographers work

This chapter turns to those aspects of geography concerned with methods and applications. It attempts to answer two main questions. First, what are the necessary skills that enable geographers to understand their world or, in other words, what are geography's tools of trade? The second question is, does the contribution made by geography to society make a difference?

Geography's key methods and skills

Throughout its history, geography has built a reputation as an empirical discipline and the practice of applied geography continues to have considerable resonance. In many ways, these qualities were inevitable. Geography started with maps as the products of exploration, discovery, and the careful recording of data, and these have been the essential tools for a wide range of human enterprises. Geography developed with the compilation of inventories of regions and places, the basic building-blocks for much of our knowledge of the Earth's surface. The so-called 'comparative method', which involved comparison of the different combinations of factors affecting different places on the Earth's surface, was often the first step to achieving a deeper understanding.

It was a short step to roles as managers of space, place, and environment. Classification and mapping of landforms, climate, biota, and soils provided, and in many respects still provide, a geographical basis not only for scientific understanding of natural environmental change and human impacts, but also for the applied fields of resource exploitation and conservation and the mitigation of human effects. In Britain, the early land-use surveys presaged the major contributions of geographers to the whole town and country planning movement. Attempts by colonial and postcolonial powers both to exploit and improve the lot of developing countries were contingent on knowledge and an understanding of the resource base.

The geographer as teacher informs children and adults of the nature of the world in which they live; of its natural order and cultural diversity. These are useful skills, but how do we identify them and trace their development in modern geography? As with most things, some have endured whilst others have diminished in significance. Perhaps more importantly, very new skills have emerged and sit within the domain of geography (Figure 20).

Fieldwork

Fieldwork is a useful initial skill to identify. It is still widely practised as a research tool and taught in the geography curriculum as an essential component of the discipline. Its origins lie in the exploration tradition: as intrepid explorers pushed their way through jungles, crossed deserts, and bridged rivers, and indeed navigated the world, they were practising and developing fieldwork skills. They were gathering information at first hand, observing landscapes and peoples as they saw them, classifying landforms and biotic species, and measuring coastlines and the elevations of mountains. As they probed into previously unknown lands and places, they had several roles. Predominantly they were West Europeans with visions of discovery, wealth, and colonies

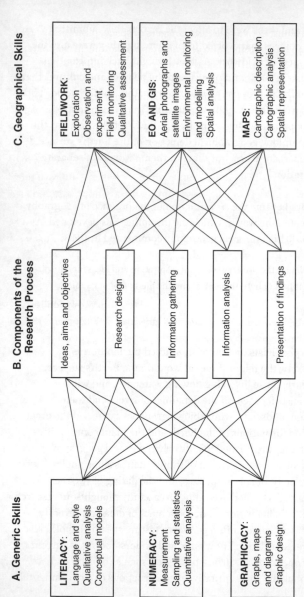

A. Generic Skills

B. Components of the Research Process

C. Geographical Skills

LITERACY:
Language and style
Qualitative analysis
Conceptual models

NUMERACY:
Measurement
Sampling and statistics
Quantitative analysis

GRAPHICACY:
Graphs, maps
and diagrams
Graphic design

Ideas, aims and objectives

Research design

Information gathering

Information analysis

Presentation of findings

FIELDWORK:
Exploration
Observation and experiment
Field monitoring
Qualitative assessment

EO AND GIS:
Aerial photographs and satellite images
Environmental monitoring and modelling
Spatial analysis

MAPS:
Cartographic description
Cartographic analysis
Spatial representation

20. Key skills of geography: (A) generic skills and (C) specifically geographical skills are shown in relation to (B) components of the research process

in mind and these were to affect the reports they submitted and the policies that evolved from them; but they were also the practitioners of fieldwork. As they brought back information about distant lands and places, and the peoples and cultures that occupied them, they were building up pictures of the world. Those pictures would be refined over time but were disseminated over a wide population. One of the by-products of fieldwork was the map, the unique geographical depiction of the Earth's surface and its characteristics, but that is a theme worthy of separate consideration.

Fieldwork became the *sine qua non* of the practice of geography; it permeated the way in which the discipline was developed. As a doyen of British physical geography commented in 1948:

> The field is the primary source of inspiration and ideas, and inspires a great part of both the matter and the method of our subject.
>
> S. W. Wooldridge,
> *The Spirit and Significance of Fieldwork* (1948)

Geomorphologists meticulously studied the landforms produced by the major forces at work on the Earth's surface, particularly those involving flowing water, ice, and wind. Hydrologists focused on the dynamics of rivers and their impact on landscape, whilst biogeographers examined the great vegetation formations of the Earth and the plant communities of which they are formed. Fieldwork in physical geography often involved measurement of things such as slope angles, water velocity, or soil properties, and the dating of surface materials, all sampled according to carefully thought-out research designs. This fieldwork produced a mass of data and several analytical outcomes, often geared to testing specific hypotheses. Measurements led to knowledge and understanding of current rates of operation of surface processes, such as erosion and deposition, and to quantitative reconstructions of environmental change.

There were detailed descriptions of the different types of physical landscapes found on the Earth's surface, but there were also many theories and models. The idea of a 'cycle of erosion', suggesting that landscapes progressed through stages of youth, maturity, and old age, was one of the earlier models. At a smaller scale, there were theories for the formation of tors in granite areas and for the plethora of strange landforms found in deserts. Climatology offers another good example where careful recording of key indicators, such as temperature and precipitation, allowed the division of the world into climatic zones that could eventually be understood in terms of the general circulation of the atmosphere, synoptic climatology, and weather systems.

Both traditional and new roles for fieldwork in physical geography can be demonstrated by the example of reconstructing Quaternary environmental change at a site on the northern coast of Mallorca in the western Mediterranean. Erosion of the coastal cliffs at Cala d'es Cans has exposed a section through a fan of deposits accumulated at the mouth of the Torrente d'es Coco, which drains a relatively small catchment in mountainous terrain (Figure 21). This example is indicative of the type of field recording that is necessary to reconstruct the sequence of events through which the landscape has passed. The field study was supplemented by several types of laboratory analyses on samples of sediments and shells; the former including Optically Stimulated Luminescence (OSL) dating, which established the timescale.

The various layers in the section reflect changing processes and fluctuations in climate. Around 140,000 years ago the area was in an interglacial, or interval between significant glaciations, experiencing environmental conditions that were a little warmer than in the present interglacial (the Holocene). At that time, the first sediment to be deposited was an aeolian (wind-deposited) sand, which later became cemented to form an aeolianite in which sand-dune bedding can be seen. Above this in the section is a sequence of layers produced under generally cooler but fluctuating

Age in years	Sediment characteristics	Geomorphic activity	Vegetation cover	Climatic interpretation
–12,000	None present	Incision of torrente gully	Reduced by humans	Moist and warm
–24,000	Fluvial and colluvial gravel and sand	Hillslope erosion; fan deposition	Low (steppe)	Dry and cold; strongly seasonal
–60,000	Palaeosol developed in loess and sand	Soil formation and landscape stability	High (woodland)	Moist and warm
–115,000	Mainly stratified fluvial (torrente) gravel	Erosion of material from hillsides; deposition of alluvial fans	Mainly low	Mainly dry and cool; strongly seasonal
–125,000	Palaeosol	Landscape stability	High	Moist and warm
–130,000	Mainly sand and gravel; marine shells present	Beach deposition (in part)	High	Moist and warm
	Cemented sand with dune bedding (aeolianite)	Aeolian deposition of sand dunes	Low	Dry and warm; on-shore winds

21. A field section through Quaternary deposits on the north coast of Mallorca: the main sedimentary units indicated on the photograph are described and interpreted in terms of changing geomorphic processes, vegetation cover, and climatic change.

climatic conditions. These layers include well-sorted fluvial gravels deposited by highly seasonal river flows of the Torrente d'es Cans and poorly sorted colluvial sediments deposited by slope processes such as debris flows. These materials appear to have been eroded when, paradoxically, conditions were somewhat drier than today but slopes within the catchment were more susceptible to erosion due to an incomplete vegetation cover. Also present in the section are buried soils (palaeosols), which signify relatively stable phases with soil development under a complete vegetation cover, which interrupted the episodes of more active landscape development. Most of the sequence in fact reflects changing Mediterranean climates through a typical interglacial-glacial cycle. The exception is the absence, at the top of the section, of the most recent sediments from the Holocene when, due to human activities on the slopes, the vegetation was degraded, soil erosion occurred, and the torrente or stream incised into the fan surface.

Fieldwork, then, has for a long time been an intimate part of hypothesis-testing, modelling, and theory development, and that quality endures. However, as some of the research problems have changed, many of the measurement techniques have improved and research designs have become more sophisticated. Measurement has evolved into environmental monitoring; observation and description now encompasses reconstruction, manipulation, and experiment; and manual fieldwork merges imperceptibly with information technology and remote sensing by satellites.

The fieldwork tradition was also strong in human geography. Perhaps the strongest single theme was the occupancy of the land, which led to the recording of land-use. The Land Utilization Survey of the United Kingdom organized by Dudley Stamp in the 1930s was one of the largest fieldwork surveys ever undertaken. 'Armies' of volunteers, mostly students, were allocated tracts of the countryside and walked around classifying land-use

according to a standard typology. The results were collated
to produce land-use maps. The exercise was repeated once.
Again, this was by a geographer, Alice Coleman, in the 1960s.
Similarly, as urban geography developed, the earliest method
was fieldwork. Information was gathered on urban land-use,
on types of buildings and the dates of their construction, on
flows of people and traffic, and on patterns of human behaviour.
Out of this, models were developed of urban growth, land-use
within cities, and classifications of urban forms. Mention should
be made of the Le Play Society, founded in 1930, and supported
by leading geographers of the day. In its expeditions, such as
those to the Balkans, it embodied all the qualities of geographical
fieldwork and this spirit survives. Finally, by way of example,
if one examines the school of cultural geography founded by
Carl Sauer at Berkeley and diffused to many parts of the world,
fieldwork was always one of its essential components. It was
one of the areas in which field sketches were widely used to
represent key indicators of cultural landscapes, whether they
were field systems, irrigation methods, or particular architectural
styles.

Although the relative importance of fieldwork to the subdiscipline
of human geography has declined with the rise of other
approaches, it is still of vital importance. Even more so than is
the case in physical geography, fieldwork in human geography
has evolved rapidly in recent years. This evolution is reflected in
the diversity and sophistication of its field methods, which range
from questionnaire surveys to unstructured in-depth interviews,
focus groups, and participant observation. Geographers studying
the neglected topics of women's lives in past periods have often
used diaries and letters as primary sources of information. The
correspondence of the wives of colonial administrators in India,
for example, has thrown considerable light on their roles in those
societies and also on their sense of remoteness and longing for
home. The diary of a Mormon woman who lived in Pine Valley,
Utah, around 1900 demonstrated the communal qualities of

life, the limited private domestic space, and the strength of the religious commitment.

For this kind of fieldwork in particular there are important ethical considerations. The interviewer needs to relate to interviewees in sensitive ways; any notions of guidance or dominance should be avoided; and the moral implications of the interaction should be carefully monitored. In other words, the researcher must always be aware of his or her own positionality. There is always a danger that aspects of that positionality, whether it be, for example, political, racial, or sexual, may influence the way in which findings are interpreted. Thus, the current importance of fieldwork to geography should not be underestimated, as aptly summarized by an American geographer:

> To me, fieldwork is the heart of geography. ... It renews and deepens our direct experience of the planet and its diversity of lands, life and cultures, immeasurably enriching the understanding of the world that is geography's core pursuit and responsibility. ... Without fieldwork, geography is second-hand reporting and armchair analysis, losing much of its involvement with the world, its original insight, its authority, its contributions for addressing local and global issues, and its reason for being.
>
> S. Stevens, 'Fieldwork as Commitment' (2001)

Maps and graphicacy

The cartographic tradition led to the nurturing of another geographical skill, which has been labelled 'graphicacy' (in contrast to literacy and numeracy). Mapmaking is a specialized profession, but generations of geography students have been taught the principles and applications of cartography to a high level of competence. Field sketches offer another example. A physical feature such as a hanging valley or system of meanders can be clearly captured by field sketches, as can field patterns or urban plans. Much of the data used by geographers, whether it is climatic, vegetational, or hydrological, or involves population migrations or retail provision, can be expressed graphically as maps, charts, or other forms of visual-spatial representation.

The central role of the map has encouraged some instruction in the science of map projections, the ways in which they are constructed and the properties that they possess. At one time in geography there was a 'great debate' over the relative merits of different types of map projections, or ways of portraying the curved surface of the Earth on a flat surface. Mercator's map projection had the quality of preserving angles exactly and showing compass directions as straight lines; it became a valuable navigational tool, but areas away from the Equator were severely distorted. The Hammond Optimal Conformal Projection minimized inaccurate representations of angles and shapes yielding a near-perfect map of a given area up to a whole hemisphere before distortions began to appear.

Graphical skills are important in both research and education and have great value in portraying the discipline of geography and its qualities. The emergence of mental maps introduced a qualitative dimension to what had always been a scientific technique. Humanistic geographers were interested in the maps that people carried in their heads. These were often lacking in precision, detail, and technical accuracy but were nevertheless important reference points for behaviour. An elderly person's mental map of the neighbourhood in which he or she lives, for example, may appear very restricted compared to that of someone younger and more mobile. Figure 22 shows two examples of mental maps derived from interviews with residents of an inner-city, terraced-row area of Cardiff, South Wales. For example A, the residents were given a list of locations (shown as dots on the diagram) and asked if they were inside or outside their neighbourhood. Three isopleths showing 90%, 60%, and 30% agreement are shown. For example B, residents were

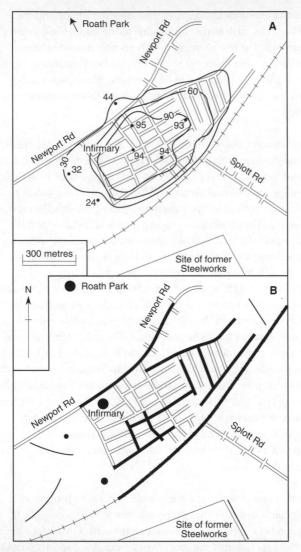

22. Mental maps of Adamsdown, an inner-city neighbourhood in Cardiff, Wales: (A) shows the degree to which residents agree on the extent of the neighbourhood; and **(B)** indicates where they locate the boundaries

asked to nominate the edges of their neighbourhood and the thicker lines, such as the railway line to the south, show greater consensus. The black circles identify specific named locations. Similarly, people chart out routes to school or local shops that avoid areas perceived as being unsafe. Such maps can be portrayed graphically by those who possess this traditional area of skill.

In many ways, the use of mental maps and images gave a thread of continuity to mapping at a time when many human geographers ceased to use maps and focused on their engagement with theory, ideology, and political awareness. Others were prompted to review their maps both in terms of their methods of construction and meanings. They continued properly to celebrate maps and the important roles that they played but were increasingly sensitive to the dangers of their iniquities, to their tendencies to enforce and encode, to the political economies in which they were embedded, and their seductive qualities. Recently observers have questioned the power of maps as not merely representing places but as creating them. In this view maps precede the real and their creative capacity has to be recognized. Thus, the significance of what maps leave out or conceal may be as important as what they include. For example, the world map first constructed in the 1830s that placed Britain centrally along the Greenwich meridian, coloured red all British colonies, and left all other land a uniform beige was a means of portraying the British Empire in all its superiority. Details and diversities were blanked out and the iconic map was designed for imperial administrators, colonial lobbyists, and settler publics.

Another dimension of the importance of maps to geography is their continued relevance in the information age, especially in relation to Geographical Information Systems (GIS). Maps are an explicit expression of the concept of geographical space and can be seen as a specifically geographical contribution to the range of methods available for understanding the world.

Numeracy

The 'quantitative revolution' in geography required the discipline to adopt an explicitly scientific approach, including numerical and statistical methods, and mathematical modelling, so 'numeracy' became another necessary skill. Its immediate impact was greatest on human geography as physical geographers were already using these methods. A new lexicon encompassing the language of statistics and its array of techniques entered geography as a whole. Terms such as random sampling, correlation, regression, tests of statistical significance, probability, multivariate analysis, and simulation became part both of research and undergraduate teaching. Correlation and regression are procedures to measure the strength and form, respectively, of the relationships between two or more sets of variables. Significance tests measure the confidence that can be placed in those relationships. Multivariate methods enable the analysis of many variables or factors simultaneously – an appropriate approach for many complex geographical data sets. Simulation is often linked to probability and is a set of techniques capable of extrapolating or projecting future trends.

This revolution forced new thinking in the discipline and a shift away from qualitative description, case studies, and the unique or idiographic, to quantitative measurements, representative samples, and nomothetic theory with its ability to generalize and predict. Geographers embraced this new approach and its associated set of analytical skills with varying levels of expertise. At one end of the spectrum were the dedicated and innovative researchers capable of interaction with statisticians, computer programmers, and mathematicians; at the other were the mass of students and practitioners more-or-less competent at basic levels of statistical analysis and the management of data sets.

An interesting demonstration of how numeracy in general, and multivariate analysis in particular, enables long-standing geographical problems to be tackled in innovative ways is provided by an investigation of the factors affecting deforestation on Pacific islands by Barry Rolett and Jared Diamond in 2004. They asked the question why, prior to European colonization, some Pacific island societies, such as those of Easter Island and Mangareva, inadvertently contributed to their own collapse by causing massive deforestation, while other islands retained forest cover and survived. Undoubtedly both different cultural responses of peoples and different susceptibilities of environments were involved. However, a comparative, multivariate analysis of nine environmental variables measured on 69 islands enabled a clear picture of which environmental factors predisposed towards deforestation and, ultimately, societal collapse rather than forest replacement and sustainability.

The extent of deforestation for agriculture, timber and fuel was quantified on a five-point scale from the accounts of early European visitors. The environmental variables measured included: rainfall and latitude, which potentially affect forest growth through climate; island age, volcanic ash fallout, and dust fallout, which relate to nutrient availability; and a number of topographic variables, which can be related to the diversity of resources, accessibility, and other effects. The strength of the relationships between these variables and their interactions, revealed by the analyses, indicated that Easter Island's fragility, predisposing it to deforestation by Polynesian people who colonized many other islands without wreaking such extreme impacts, can be attributed to a combination of a relatively unfavourable climate, nutrients and topography. In other words, Easter Island society did not collapse because the people were unusually improvident but because they occupied a particularly fragile environment and were unable or unwilling to adapt their farming methods to suit its needs.

The introduction of quantitative methods into geography as a generic skill had many repercussions. The development of numerical models was particularly important. In physical geography, it allowed the development of process models that had a much stronger scientific basis than descriptive models such as the Davisian cycle of erosion (see box in Chapter 2). In human geography, the models had been distinctly sparse but the models of city structure and growth were worthy of that description, though they were derived from outside geography. The kind of advance made possible is exemplified by Torsten Hägerstrand's work on diffusion, which employed sophisticated yet transparent probability theory to demonstrate the spread of innovations over time and space. Figure 23 offers one of the early examples from his studies in rural Sweden and is concerned with the gradual acceptance of subsidies by farmers over time. Diagram A shows the simulation grid made up of cells for which probability scores based, for example, on distance from the source of subsidies, can be calculated. This grid is superimposed over the actual area in B, on which is shown the actual diffusion of subsidies after three years. Such spatial models are also used in medical geography to analyse the spread of diseases and in population geography to demonstrate patterns of migration over time.

It is probably true to say that only a minority of human geographers currently retain an interest in numerical analysis but this state of affairs has its critics. For example:

> Geography is losing its way precisely because so many if its practitioners have retreated from the quest of creating robust, defensible generalisations about spatial patterns and processes.
>
> P. A. Longley and M. J. Barnsley,
> 'The Potential of Geographical Information Systems' (2004)

There is, however, one field of geography where numerical skills have found a new home and are forging ahead; that is the science of Geographical Information Systems (GIS).

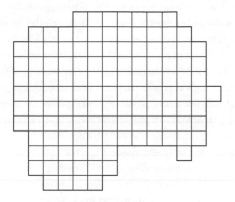

A Simulation grid

B Diffusion after 3 years

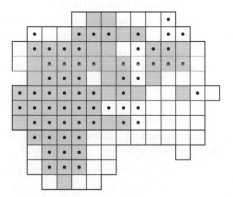

• Actual take-up of
   subsidy by farmers

▨ Predicted take-up

**23. A spatial diffusion model: (A) the simulation grid over the Swedish
study area; (B) actual and predicted take-up of subsidy by farmers
after three years**

Geographical Information Systems (GIS)

GIS science is a major modern skill of the discipline of geography. It has developed with increasing diversity into various forms of sophisticated mapping combined with quantitative spatial analysis. GIS data comprise digital representations of phenomena found on the Earth's surface. These may be landforms, field boundaries, vegetation types, buildings, or a host of other features that can be referenced to geographical coordinates. Once this data has been collected, GIS software, such as MAPINFO, allows a range of analyses and interpretations. GIS science has a well-defined and developing set of scientific principles, practices, and theories, and the methodology has proved to have very considerable application. Global sales of GIS facilities and services exceed $7 billion and find markets throughout public services such as local government and the police and in many private sector areas such as financial services and retailing. GIS applications are problem centred and address long-standing research questions in geography such as urban growth and land-use change as well as newer challenges such as crime profiling, where many police forces routinely add a location code to criminal events such as burglaries or homicides as a first step towards forms of GIS analysis. The example shown in Figure 24 is from human geography and demonstrates the way in which GIS can present data in different ways. This map of the countries of the world shows the varying distribution of wealth, based upon the Gross National Product (GDP). The territorial extent of each nation state is distorted to reflect its relative prosperity. Countries such as the United States and most European states have a greatly exaggerated size whereas countries in Africa and Latin America are minimized.

GIS is one half of a duality of which Earth Observation (EO), an alternative term for satellite remote sensing, is the other partner. EO comprises a set of instruments or sensors, their carriers, aircraft or satellites, and the data processing techniques that

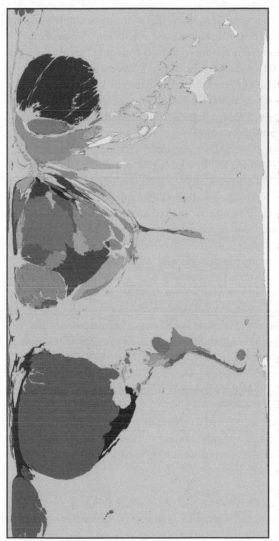

24. The distribution of wealth plotted by WORLDMAPPER: the application of a Geographical Information System. The area of each country is proportional to its gross domestic product (GDP), which emphasizes the gross disparities between the developed and developing worlds

can be used to gather information about the Earth's surface from a distant location. EO provides an important source of data for GIS and working in tandem they have permeated a large number of now taken-for-granted applications in the real world. Satellite navigation in cars is one example of everyday use as is the increasing popularity of Google Earth (the accessible set of satellite images available on the internet). At more demanding scientific levels, the technology is ideally placed to monitor environmental change globally, over large areas or in remote locations. Climatic change, the diminution of glaciers and sea ice in the polar regions, the spread of desertification in sub-Saharan Africa, soil degradation in the American Mid-West, and the clearance of tropical rainforests in the Amazon Basin provide spectacular and important applications of this type.

Desiccation of the Aral Sea over recent decades provides a striking example of the use of satellite images in monitoring environmental change (Figure 25). In 1960, the Aral Sea was the fourth largest inland water body on Earth, with an area equal to the combined area of France, Germany, Spain, and the United Kingdom. It was mismanagement of water abstraction from the rivers Amudarya and Syrdarya for irrigated agriculture, the expansion of which was particularly rapid between 1976 and 1988, that led to this desiccation. Desiccation of the Aral Sea has led to decline in the groundwater table, salinization, expansion of halophytic (salt-tolerant) vegetation, deflation of the exposed sea-bed and aeolian deposition on surrounding land surfaces from salt and sand storms. Other knock-on effects contributing to desertification of large areas of the Aral Sea drainage basin include waterlogging and secondary salinization following irrigation, and chemical pollution of soils, groundwater, rivers, and the Aral Sea itself from toxic chemicals used in cotton production. The Aral Sea case is a classic example of a 'creeping environmental problem', induced by human action, ultimately leading to an 'environmental disaster' and the mass emigration of people from the affected region. It is only since

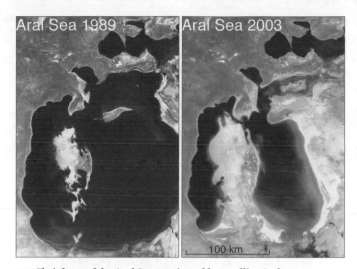

25. Shrinkage of the Aral Sea monitored by satellite. Left: July–September 1989; Landsat mosaic at 500m resolution. Right: 12 August 2003; Aqua MODIS scene at 500m resolution

1997 that there has been a partial reversal of trends in relation to the northern part of the Aral Sea (the so-called Small Aral), whereas the processes may be irreversible in relation to the Large Aral.

It is worth remembering that the precursor of EO, aerial photography, has had a presence in geographical research and in the geographical curriculum for a long period of time, but the power of the new systems and satellite imagery has moved us to new dimensions. GIS and EO are strongly technological in their make-up but around that technology have developed powerful models, processes for analysing data, and methods of interpretation. They are both massively effective information-gathering procedures, but the value of this information is only realized when good interpretation is able to make the most effective use of it.

It is salutary to note that the largest annual meeting in the world devoted to an aspect of geography is not organized by academic geographers but by a private GIS company, the Environmental Systems Research Institute. Ownership of the GIS/EO field threatens to move outside the discipline of geography to specialized institutes, possibly to engineering departments of universities and to government agencies. In many ways, this is a healthy sign of a success story, and geography has no monopoly over this increasingly vast enterprise. The important thing is to retain a strong presence of GIS and EO within the discipline of geography where it surely belongs. It has to be acknowledged, however, that GIS and EO sit less comfortably in a discipline that has experienced such radical changes since the decades of the quantitative revolution when there were few challenges to the idea of geography as a science.

Literacy

Perhaps the final skill expected of geographers is 'literacy'. That assumption has always been there and was present in the trilogy of 'books, benches, and boots' that used to be embedded in the minds of all geography undergraduates. As 'benches' emphasized laboratory work, practical classes, and cartographic and statistical skills, and 'boots' hammered home the importance of fieldwork and the 'field' in general, so 'books' drew students back to the fundamental need to master the literature of their subject and be able themselves to write it. Literacy of course is fundamental to all academic disciplines and has no claims to be counted as a special skill of geographers. Similarly, numeracy is widely used in social sciences such as economics and psychology. What has changed in recent years, and is clearly linked with the rise of the new cultural geography, is that geographers, especially human geographers, have been drawn into and are expected to be familiar with areas of literature beyond their former experience.

To some extent, this is to be expected as the research frontiers push forward, but with the new cultural geography new dimensions have been reached. Critical theory, postmodernism, and post-structuralism have drawn geographers into the literature of philosophy *per se*, well beyond the philosophy of science. This has to be a positive trend but begs the question of the amount of literature geographers can get their heads around? Do they expand into a wider field or retain their focused, disciplinary niche? David Harvey is sceptical about the continual importing of new thinkers, theorists, and theories into 'the grand parade of external interlocutors as to what geography might and should be about'.

There are additional skills to be derived from these trends. The new literature is shared with a burgeoning field of cultural and media studies; modern society looks for skills and awareness in these areas and many of the study themes have strong communality. The works of Jean Baudrillard, who writes powerfully about imagery and the ways in which landscape and society can be 'read' from the 'texts' that are portrayed through the likes of posters and billboards, and Ferdinand de Saussure and Jacques Derrida, who focus on signs, icons, and the study of semiotics, provide examples of 'imports' from French intellectuals. Growing interest in discourse analysis and non-representational theory (see box in Chapter 3) adds new dimensions.

The new 'subjects' for interpretation include film, works of art, music, dance, and theatre, all of which carry messages about the societies in which we live. As Sarah Whatmore, a British geographer, has commented, the world speaks through many voices and we may rely too much on the spoken and written word. Novelists such as Dickens and Steinbeck wrote works of fiction, but their books throw light on the contemporary societies and places in which their stories were set. Painters such as Renoir and Rembrandt leave us with depictions of people and settings from the narrow segments of European society that they chose, or were sometimes commissioned, to depict. Dance and theatre are diverse throughout the world, but carry messages about the cultural settings from which they emerged; music such as the 'blues' emerged from very particular backgrounds. All of this literature and 'directions' for study add to the skills of human geographers, or at least increase their awareness of the diverse sources of information and theories that exist. Whereas physical geographers have less need to access this increasingly diverse literature, there is growing awareness of its significance, particularly among those involved in collaborative projects.

Applied geography

Applied geography proclaims the importance of the empirical tradition and its relevance to problems of the real world. In some areas of human geography in particular, this tradition and the theme of applied geography has been discarded or at least relegated to a less prominent position. Yet applied geography as a term brings together these many skills that we have identified and explored: and offers the opportunity for practical relevance. Furthermore, it is not really an option to concentrate on either theory development or applications: there are synergies between them. As a practice it would be discarded at our peril, and in reality it continues to make significant contributions.

Applied geography involves the utilization of acquired skills and knowledge to address problems and issues of the real world. It can take several forms:

- *As input to policy-making and the setting of research agendas.* It important to engage with international research agencies and development practice, and this engagement is made through governments and non-governmental organizations (NGOs) with far-reaching consequences on research and influence on policy. There are opportunities not just to respond to externally defined agendas but to help shape them in the first place. The Danum Research Centre in Sabah, Malaysia, for example, monitors the causes and impacts of deforestation in the Tropics. Recent experience of tsunamis in Asia, hurricanes in the Caribbean, and flooding in Britain has stimulated the need for greater awareness, early warning systems, and preventative measures.

- *As direct involvement of geographers on key committees and working parties.* The ability to form research agendas is considerably enhanced if geographers are members of key committees in both the public and private domains. Roles as advisors to government and lead roles on front-line working parties are especially influential. The British geographer Sir Peter Hall has had considerable influence in the development of urban policies. He has served on the South-East England Regional Planning Committee, on the Prime Minister's Urban Task Force, and was an advisor on trunk road developments in the UK. He was also Special Advisor to the Secretary of State for the Environment 1991 to 1994 and has testified before the US Congress on urban policy. He is regarded, moreover, as the founder of the Enterprise Zone concept, which has been adopted worldwide. The Enterprise Zone was designed to encourage private economic development in designated areas. Within this 'zone' investing companies enjoyed tax concessions and relaxed planning regulations. At a different level, there is an excellent record of the involvement of geographers in UK public bodies such as National Parks committees, Sports Councils, and Crime Prevention panels.

- *As parts of an interdisciplinary team to tackle major global or regional problems.* Increasingly governments are becoming aware of the need to turn to science to understand and tackle issues such as global warming and environmental change. Geographers need to obtain presence on these major projects and there is ample evidence for success in this context. Groups dedicated to understanding the current and future changes in the Greenland and Antarctic ice sheets involve geographers; many of the major European consortia funded to study key issues such as environmental change over the last millennium and sustainability involve geographers.

- *As a specific contract or consultancy to address and offer solutions to a current issue.* There is a long record of one-off projects whereby geographers have been commissioned to undertake specific tasks of evidence-based research. These are sometimes supported by Funding Councils but also by the private sector. Studies of retail change, whether it be store locations, market definitions, or consumer behaviour, offers one example. Optimal location projects might involve airports, marinas, hospitals, and a range of public utilities. These types of applied geography often do not push at the research frontiers but have practical value.

- *As a by-product of research that may have some other conceptual purposes but from which practical applications emerge.* This is the most common source of applied geography and the most diverse. So-called 'blue skies' research (with no direct applications foreseen in advance) may stimulate an interest in an evidence-based sideline that may later yield applications. The thrust in human geography towards discourse analysis and non-representational theory is at times obscure but its main purpose is to achieve a better understanding of the societies in which we live. Historical geography examines the past in a variety of ways that often, intentionally or unintentionally, throws light on contemporary issues. The ways in which young migrants, especially women, to Swiss cities in the early 20th century were controlled in an attempt to maintain middle-class standards of social order has some resonance for modern refugees (see box). Again, a detailed study of ghostly heritage of a district in Singapore that once held female labour collectives in the so-called 'death houses', shows how the image has persisted in the modern city and affects the way in which the area is regarded (see box). These examples show how a great deal of scholarship in geography touches upon the real world and has something relevant to offer.

Women migrants into European cities

During the later 19th and earlier 20th centuries, the migration of young women into European cities was regarded as problematic by established members of middle-class society. There were fears for their morality that translated into the regulation of sexuality and gender. Cities were seen as dangerous places for the trafficking of girls for immoral purposes and societies were established to counteract this perceived threat. The International Association of Friends of Young Women was established in Switzerland in 1877 and eventually claimed to have helped 35,000 young women. Activities were not confined to Switzerland, and in Berlin, for example, over 80,000 women were helped. Railway stations were seen as having key roles as gateways into the new ways of life, and the train station assistants were so placed to guide, give advice to, and help new migrants. Middle-class mores revolved around Christianity, motherhood, and altruism, and these were practised as a means of maintaining standards of behaviour and social order.

The above listing of types of applied geography has included brief descriptions of the kinds of activities that fall within this category. A fuller example is offered by research into natural hazards, a theme that involves both physical and human geographers, serves as an integrated example within geography, and is also interdisciplinary as it links with the skills of geophysicists, engineers, psychologists, and communications experts. Its established presence in geography demonstrates the necessary synthesis of methods and ideas.

A natural hazard has two main components, the physical event or process that often occurs cataclysmically and the vulnerability of people located in its area of impact. The physical event can be

Ghostly geographies of Singapore

One fragment of the urban landscape of Singapore has, very uncharacteristically, remained empty and unkempt over a long period of time. It was an excluded site, in a city renowned for its economy of space and order, right through until 2006. The site has a history connected with the influx of female migrants from Guangdong in China to provide cheap labour. They occupied what became known as the 'death houses' because of burial collectives and the conditions under which they and many old people in the area lived. The buildings were bulldozed in 1969, but the reluctance to develop the site reflected the heritage of burial places and haunting. There is a more general issue of death and haunting that permeates Singaporean landscapes and relates to a programme to remove graves and transfer remains to central locations. Offerings to the dead still appear in many parts of the city, and the fear of haunting can affect both development and the value of property. This particular site remains an urban wound, something of a funereal landscape, where the heritage of the past still affects a modern city. The Hungry Ghost Festival in Singapore bears testimony to a dimension that contested the use of space in the city.

a range of things that include earthquakes, volcanic eruptions, floods, storms, and landslides. The event becomes a natural hazard if it has an impact upon people or property. For example, the Great Alaskan Earthquake of 1964 displaced 29 million cubic metres of rock, and slid down the Sherman valley at speeds of up to 180 kilometres an hour, but had no impact on people. In contrast, the Aberfan slide of 1966 in Wales had 1% of that volume and travelled one-twelfth of the distance at one-twentieth of the speed, but killed 144 people. The latter was a natural hazard that led to a disaster. We live on an unquiet Earth, as the 2004

Indian Ocean tsunami, which killed around 230,000 people, and Hurricane Katrina, which in 2005 killed 1,836 people and left 80% of New Orleans under water, testify.

The roles of physical and human geographers in studying natural hazards are different but complementary. Physical geographers need to understand, monitor, and predict the actual physical event; human geographers need to understand how the dangers of natural hazards are perceived and whether they influence decision-making and behaviour. Studies of post-tsunami Sri Lanka show how fear of recurrence has become a political tool to implement a range of security measures, including a buffer zone that has caused discontent among those affected by its location. There is a need to set up preventive measures and mitigating policies, including zoning regulations, which require explicit recognition of the spatial properties of the hazard, but consultation and collaboration with local residents is crucial.

The record of applied geography's intervention has not been uniformly successful. The physical events, such as a volcanic eruption or earthquake, can be monitored but prediction is by no means a precise science, the margins of error are wide. The experience of human behaviour is that whatever advice is given, people will build on flood plains and return to settlements close to volcanoes. The lack of imperatives in the human response may in part be due to the imprecision with which the physical events can be predicted, but it also stems from cultural norms, political contexts, and patterns of behaviour that need to be addressed. Physical and human geographers clearly need to be involved together in hazard research, but there is some evidence that physical geographers work more comfortably with geophysicists and engineers, who speak the same language of science. There is also evidence that even when the science of the event is understood, a natural disaster will still occur where that evidence has not been successfully communicated. In the Nevada del Ruiz volcanic eruption of 1958, there was an almost complete hazard assessment in place and careful monitoring, but still 23,000 died from the mudflows. There was a failure to communicate the dangers with sufficient clarity to those at risk. Gilbert White's classic model of hazard research, developed from his work on floods, moves through the three steps of physical event, human vulnerability, and human consequences of disaster. An alternative model is to place human vulnerability as the first step on the grounds that the population at risk defines the natural hazard.

Natural hazards research is well established. It has made significant contributions to many kinds of situations including flood plains and now coastal inundations in an era of global warming. It is a very clear example of the need to apply geographical skills in an interdisciplinary context in which both physical and social sciences have key roles to play.

Chapter 6
Geography's present and future

Early explorers, mapmakers, and regional geographers from the last century, and even quantitative and Marxist geographers of the late 20th century, would find much that was unfamiliar in the modern practice of geography. Certainly, many core values remain but the manner of their study would bemuse, and some of the new additions to human geography in particular would defy, their imaginations. In many ways, of course, that is to be expected. All disciplines move forward in time and change in the process.

Having reviewed the ongoing trends, some of the newest directions that geography is taking can be considered, its present state can be reviewed, and its future contemplated. This concluding chapter has two parts. The first part sets out some fairly detailed examples of innovative modern research in geography, which emphasizes the forward momentum. In the second part, in a 'manifesto for future geography', we seek answers to two main questions: (1) where does geography stand now as a unified discipline; and (2) can that unity survive in the face of increasing diversification and continuing change?

Some modern faces of geography

Geography has many modern faces, some of them surprising and unexpected. This final chapter begins with vignettes that

26 Al Capone in May 1932, on his way to the Federal Penitentiary in Atlanta, Georgia, where he would start serving an 11-year sentence

exemplify some of these. Two examples are taken from human geography and two from physical geography.

The geography of crime

One might perhaps not expect the face of one of the world's most notorious criminals (Figure 26) to grace a book on geography but, over the last two decades, a strong interest has developed in the geography of crime and what has become known as 'environmental criminology'. Studies often begin with spatial patterns – there are clear concentrations of crimes in space, and there are crime areas characterized both by large numbers of criminal events and the residences of many known offenders. The Chicago where Al Capone thrived in the 1920s and 1930s became an early laboratory for studies of this kind, and spatial descriptors such as zones, gradients, and delinquency areas emerged from these studies. Chicago was divided into clear gangland territories where different groups, often ethnic, held sway in the years of 'prohibition' in particular. Spatial patterns were only a starting point as it soon became clear that offences and offenders had different geographies aligned to opportunities in the environment or targets, on the one hand, and the social condition of the neighbourhoods, on the other. There are links between poverty, deprivation, and many offences: crimes of violence typify places of conflux and entertainment; white-collar and corporate crimes have different geographies. Urban geographers have explored the incidence of specific offences such as burglary and have shown which kinds of neighbourhoods are the most vulnerable. There are several hypotheses to identify most vulnerable areas:

- The 'offender-residence' hypothesis suggests that places where many offenders live are vulnerable (burglars tend not to travel far to offend).
- The 'border-zone' hypothesis suggests that the edges of neighbourhoods are the most vulnerable.

- The 'local social-control' hypothesis suggests that neighbourhoods with a strong sense of place and high social interaction are less vulnerable.

- The 'area-variability' hypothesis suggests that mixed residential areas with high levels of transience are vulnerable.

These are hypotheses for which some evidence is available but this is rarely conclusive. The local social-interaction hypothesis is closely aligned to the policy of 'neighbourhood watch' where residents, in collaboration with crime prevention officers, look out for each other's property in the fight against crime.

Along with crime itself, there is strong evidence of the importance of fear of crime. Vulnerable people, such as the elderly and women with young children, are often reluctant to move around certain parts of the cities and areas such as open spaces after dark are avoided. Research has moved on to examine the roles of the police in shaping the geographies of crime. A well-known study of vice areas in San Francisco showed that the spatial shifts in these over time were products of the police and the criminal justice system in changing the rules of behaviour rather than the 'offenders' *per se*. The police carry mental maps of the cities in which they work and these can affect the forms of policing and the responses to crime. Problem estates are the products of both those who live there and of the gatekeepers and agencies that allocate the housing, set the standards, and apply the rules.

Such geographies of crime lend themselves to interpretation by the different approaches in geography. A spatial analysis approach, for example, would map and correlate; a Marxist would be more interested in the ways in which the unequal distribution of wealth and opportunities creates crime in the first place; a behavioural geographer might study the decision-making process a burglar follows or the mental maps he or she holds of the target city. A postmodern approach would question the discourse of a criminal justice system that labels an act as deviant in the first place; crime is a social definition. Geography also has practical value in coping with crime. Most police forces are using GIS and create maps of offences and crime scenes. Forensic psychologists have used basic techniques such as centrographic analysis, which generalizes upon sites at which offences occur, to profile the locations where serial offenders may be found. This is one modern face of geography where both old and new methods and intellectual traditions have been applied to a fresh and different subject area.

Geographical meanings in literature and film

A second vignette can be used to demonstrate the changing meanings that underpin the subject matter of human geography. There are well-known attempts to employ fictional literature as a means of gaining insight into the places where their novels were set. Charles Dickens, for example, throws parts of London into sharp perspective and has a great deal to say about the social condition of the people; Jane Austen sketches the lifestyles of genteel, upper-class rural families at the end of the 18th century; and Upton Sinclair portrayed the appalling conditions under which the poorer people of Chicago lived in the early 19th century. Similarly, works of art can be used to interpret a view of landscapes. Constable's paintings of rural England represent tranquillity and continuity; the Impressionists, such as Monet, depicted the leisured classes of French society in the settings, rural and coastal, where they resided or played. Sources of this kind have always been used with caution. Novels, for example, are works of fiction and authors are not necessarily constrained by adherence to real facts. Postmodernism opens up a more critical view that applies not just to literature but also to narrative history. The essential argument is that all 'facts' are not real but must be seen relative to the writer and the values that he or she holds. Jane Austen, for example, belongs to the stratum of society about which she writes; her knowledge of other parts of the world and of its societies is closely circumscribed.

There has also been an interest in film as a means of interpreting society and place, but in this example a British geographer, David Clarke, closely interweaves his interpretation with an interest in critical theory. He sees the theoretical stance that he takes as an essential lens for understanding how the film works. Patrick Keiller's *The City of the Future* is a film, based on an adventure story, in which an individual, who becomes the narrator, travels back in time to seek a historical figure, Dr Karl Peters, who was German and wrote a book called *England and the English* in 1904. Peters, who had worked for Germany in colonial East Africa, was thought by some to be an inspiration for the character of Kurtz that Joseph Conrad introduced in his novel *Heart of Darkness*. The plot in *The City of the Future* was for the time traveller to intercept Peters in his journey around England and in so doing change the course of history to avert subsequent disasters such as World War I.

The images used in the film explore the contrasts between the familiarity of the old city fabric, the strangeness of the past, and the newness of present-day experience. Time travel is used as a narrative device against which the plot unfolds. Keiller follows the line that Britain was 'peculiarly capitalist', with a brand of London-based colonial capitalism that ensured differences from the rest of Europe. He uses the images to show that built environments of the British cities against all the expectations of modernism have changed at a snail's pace compared with everything else. His images of the past are strangely familiar, yet everything else in the environments of the city bears witness to the fact that they belong to a different age. There is a contrast between the familiar-looking landscapes and the unfamiliarity of the society glimpsed within them (Figure 27). Keiller sees an *unheimlich* effect, 'a profound disjuncture between society and space with the loss of both the humane city and its utopic future'. *The City of the Future* is presented as an adventure story, but its journey into history is an attempt to recuperate the past in an effort to redeem the future. *The City of the Future* rests on the

27. Oxford Street in the 1920s: the basic form, or morphology, of this famous London street is clear and reasonably constant, but the people and signs of technology indicate another age

belief that the solution to preventing the crisis over the loss of a future may be found in the past. The time traveller fails in his search for Peters and the mission is unfulfilled: *The City of the Future* concludes that the world from which we want to escape is that in which we are most involved and the mission to de-create or reconstruct the real or what actually happened is doomed to failure.

The film explores the intentionality of the film-maker and his view of the world. It engages with critical theory as it explores the pervading influence of romanticism. The defining quality of romanticism is seen as its assertion that *being* should yield itself to *meaning*. Yet the error of Romanticism is its belief that the human subject can form its own identity when in fact it is formed by the *Other* (or those who observe his or her activities). The subject cannot dominate the gaze of the *Other*; it is subservient to that gaze. It is this reliance on romantic precepts that diminishes the interpretation of past and future that Keiller wishes to portray. This vignette illustrates the widening engagement of some human geographers with literature ranging from Lacanian psychoanalysis to critiques of modernism. At the same time, it addresses the nature of urban landscapes and the people who occupy them.

These examples are specific and can be illustrated by particular case studies. Many of the new approaches to human geography have more general aims and seek to question previous interpretations. The box offers a few summary statements on the development of 'hybrid geographies' that fall into this last category. Sarah Whatmore describes her research in this field as focusing on the relations between people and the living world, and the spatial habits of thought that inform the ways in which these relations are imagined and practised in the conduct of science, governance, and everyday life.

Geo-ecological studies on glacier forelands

Glacier forelands are the recently deglaciated zones in front of retreating glaciers (Figure 28). Since attaining their maximum extension of the 'Little Ice Age', most glaciers have been generally retreating for several centuries, exposing new land that has begun to evolve. These special places provide physical geographers and others with an opportunity to investigate the development of vegetation, soils, landforms, and other aspects of the landscape. Glacier forelands can be viewed as field laboratories in which a 'natural experiment' is unfolding: close to the glacier, the landscape is freshly exposed and devoid of life; farther away, it has been exposed for longer, plants are colonizing, soils are developing, and slopes are more stable.

Like other types of experiment, glacier forelands present a situation where at least some of the complexities of nature are simplified. The geo-ecosystem is relatively simple, its history

Hybrid geographies

One of the most central questions in geography, the relationship between Nature and Culture, is addressed in the British geographer Sarah Whatmore's study of hybrid geographies. The central argument is that nature and culture are not antitheses but are closely interconnected. These various and intimate connections are best studied by investigating the attachments, skills, and intensities of differently embodied lives rather than by reference to major academic issues or corporate management. 'The politics of global ecology are necessarily more plural and partial than a global vision that maps a universal subject, the "we" of humanity onto a finite terrain.'

The theme of the relation between human and non-human is pursued in some of the case studies that make up the book. A section on 'The Wild' questions the ways in which animals are managed in nature. A designation of Wild, or wilderness, does not seem to have served animals well; they are caught up in networks of regulation and wildlife management that serve human rather than non-human interests. On the example of genetically modified (GM) foods, Whatmore sees 'food scares' as the outcomes of the threadbare trust that exists between growers, suppliers, and consumers. The plea is for a stronger voice for those who understand agri-food production and consumption as a science in the chain of decisions that leads from field to plate.

of development is short, the spatial scale of the landscape is manageable, and, most importantly, the age of the terrain is known or can be dated. This last aspect is particularly important and has led to the chronosequence concept: the idea that distance from the glacier front represents the age and hence the stage of development of the landscape. In other words, space can be

28. **Storbreen glacier foreland, Jotunheimen, Norway: a landscape unit and geo-ecological system. The isochrones (lines of equal land-surface age) show the retreat of the glacier since the 'Little Ice Age' maximum of the mid-18th century**

regarded as a substitute for time in the glacier-foreland landscape. This enables the study of the changing landscape over at least centuries, an otherwise impossibly long period of time to directly observe or monitor change. Precise dates of deglacierization of the land surface (Figure 28) permit the inference of precise rates of change in the landscape.

What have biogeographers found out about vegetation from studying these landscapes? By investigating patterns within and between forelands, the rates and trajectories of vegetation succession through time have been related to local habitats and regional environmental conditions. The simple notion of a single pathway of succession towards a stable ('climax') state has been refuted, with diverging pathways leading to different mature states controlled by environmental gradients such as altitude, moisture, length of snow-lie, and frost disturbance of soils. These results amount to a geo-ecological theory of succession that has applications in the field of land reclamation and rehabilitation. Universal solutions to the damage that human activities inflict on geo-ecosystems are unlikely: different, or at least modified, solutions are necessary depending on geographical considerations, such as the particular environment and position in the landscape.

The uniquely geographical contribution has been threefold. First, a geographical perspective has led to thorough knowledge of the nature and extent of spatial variation on the glacier foreland. Second, this has led to a more realistic chronosequence concept: distance from the glacier and terrain age are not the only variables to be considered in explaining these landscapes. Third, the geo-ecological approach includes a holistic appreciation of the interactions between physical and biological processes in the development of the various landscape elements. Similar contributions have been made by zoogeographers, soil geographers, and geomorphologists to knowledge and understanding of topics such as insect succession, soil development, rock weathering rates, the development of periglacial patterned ground, and moraine-ridge formation by the glaciers themselves.

The geography of global warming

Of the many new topics to be taken on board by physical geography in recent years, global warming has undoubtedly had the most far-reaching effects on what geographers do. The latest report of the Inter-Governmental Panel on Climate Change (IPCC) concluded that global mean surface temperature rose by about 0.76 degrees Centigrade over the last 100 years and, if atmospheric carbon dioxide concentrations double, a further rise of about 3.0 (likely range 2.0 to 4.5) degrees Centigrade can be expected to follow. Past uncertainties about the rate and underlying human cause of this global rise in temperature have now been largely removed. Its likely worldwide impacts on natural environments and on people are much clearer, and there is increasing support, for example from the Stern Report, for an economic case for taking action to reduce carbon emissions. Furthermore, it is also clear that a global response is required to this global environmental problem.

Why are physical geographers in particular taking an interest, and how do they contribute to solving this environmental problem? The answer is that global warming has many geographical aspects. These can be illustrated with reference to several different dimensions of the global warming problem: namely, detection, prediction, impacts, and mitigation.

Detection of the problem in the first place requires measurements of air temperature taken from many parts of the world. Similarly, geographical variations in temperature over the Earth's land surfaces and over the oceans must be taken into account in calculating accurate estimates of globally averaged temperatures and to refine our knowledge of the rate of global warming. The term 'global warming' also tends to hide the fact that different

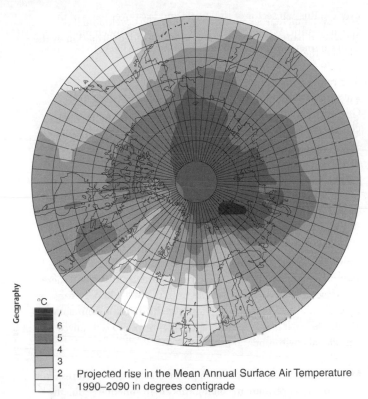

°C
7
6
5
4
3
2
1

Projected rise in the Mean Annual Surface Air Temperature
1990–2090 in degrees centigrade

**29. Arctic warming: mean annual air temperature rise projected
for northern latitudes between 1990 and 2090 according to the
Arctic Climate Impact Assessment (2004). Darker shading indicates
relatively great temperature changes in the Arctic**

parts of the globe behave differently. In the polar regions, for
example, the warming trend has been greater than elsewhere;
and according to recent estimates, Arctic warming over the next
century may be double the global average. Figure 29 shows the
projected surface air temperature rise for northern latitudes
from AD 1990 to 2090 according to the Arctic Climate Impact
Assessment (ACIA). The temperature rise is likely to be greatest

over the Russian sector of the Arctic and least over parts of the North Atlantic. Winters are likely to be affected much more than summers, while precipitation changes are likely to be more variable.

Physical geographers play an important role in testing the General Circulation Models (GCMs) that simulate the Earth's climate system and predict the likely course of future climate. While few geographers have the physical and mathematical skills to design GCMs, they are widely involved in the multidisciplinary effort to test these models using our knowledge of past climates. Because GCMs are predicting the future, they cannot be tested by conventional observation and experiment: but they can be tested by seeing whether they are able to predict what has already happened. This is why reconstructions of past climates by physical geographers and others are so important. Such palaeo-climatic reconstructions are made using evidence from many different sources (such as ice cores, peat bogs, lake sediments, and tree rings) collected from many different environments. The accuracy with which the models can predict palaeo-climates that differ from present-day climates provides a measure of the confidence we can have in their ability to predict future climatic conditions.

The impacts of global warming also vary in different parts of the world. Natural environmental systems differ in their sensitivity to rising temperatures, as do human systems. An example of a highly sensitive natural system is provided by the African Sahel, where relatively small changes in annual or seasonal temperature can greatly influence moisture availability, vegetation growth, and dependent cropping and grazing economies. Regions with lower temperatures and/or more frequent rains provide a starting point that is less vulnerable to drought. A very different example is provided by the impact of rising temperatures on the glacial and periglacial environments in the European Alps. There, warming has led to rapidly diminishing glaciers, an increase in the frequency of debris flows following the thawing of the underlying permafrost, and concern for the long-term viability of the hydro-electric power installations that utilize summer meltwater. In the Arctic, additional implications include the extent of pack ice, polar bear survival, and the future of cold-water fisheries.

Finally, a one-fits-all solution is unlikely to apply to mitigation of the effects of global warming on people. Societies differ in their vulnerabilities, and human geographers are well placed to investigate these. Some societies are more able than others to adapt to changing environmental conditions or will choose to respond differently. In general, people in the poorest countries are more vulnerable than those in rich countries and are also less able to adopt expensive technological solutions. This is a different aspect of mitigation from seeking to reduce the rate of global warming or taking global action to prevent the rise in global temperatures exceeding an upper limit, both of which may require different policies for developed and developing countries.

A manifesto for future geography

Geography, at the beginning of the 21st century, has indeed spread its net wide. On the one hand, there are the core unifying concepts and skills that remain important to the discipline as a whole; on the other, there are the tensions between the subdisciplines within which specialities have multiplied apace. The contrast with geography as it was practised from the later 19th to mid-20th centuries is very clear and leads to the question: Where does geography stand now as a unified discipline?

Present geography's strengths

The discipline has a significant number of strengths and opportunities that in many ways are increasing. There has never been a greater need for geographical knowledge and understanding. Whether it is in relation to local or global environmental problems, or local or international conflicts, this is everywhere apparent. The general level of geographical literacy, or graphicacy, with its emphasis on visual-spatial skills, needs to be raised for wealth creation, for preserving and enhancing the quality of life, for ensuring sustainability of the Earth and its peoples, for responsible citizenship, and for leadership at local, national, and international levels. This presents not only research opportunities, but also educational opportunities in preparing people to meet the physical and human challenges of an ever more crowded, unequal world with an ever more competitive global economy. Never has the phrase 'geography matters' had more resonance than in the present day.

The breadth of the discipline of geography is a major source of strength. Physical geographers contribute knowledge and understanding as natural environmental scientists, while human geographers play distinctive and important roles as social scientists and as social and cultural theorists. Alongside this, those who work on the integrated themes of regional, historical, human–environment interaction, global change, or landscape geography continue the holistic traditions of geography and exemplify its bridging role. In terms of research, therefore, the specialist contributions of physical and human geographers are complemented by the concerns of the integrated geographers with the broader questions and with synthesis. Students of geography at university and school pupils benefit considerably from this broad remit. Geography is of immense personal educational value, preparing flexible graduates with the breadth of knowledge and the wide-ranging skills – numeracy, literacy, and graphicacy – necessary for a wide variety of careers.

The core geographical concepts of space, place, and environment are more relevant than ever to understanding the world. Geography now possesses a much greater knowledge base and a better-developed battery of methods, including its own specialist geographical techniques associated with maps, EO, and GIS, than at any previous time in its history. Neither the complexity of the Earth's surface nor the short history of the discipline can be regarded as the weaknesses they once were. Geography is now well prepared to perform its mission.

Environmental concerns present geography with obvious opportunities. First, there are the opportunities associated with the need to understand the biophysical environment itself. Past, present, and future patterns, processes, and changes in the biophysical environment need to be reconstructed, measured, monitored, modelled, mapped, and predicted by physical geographers. Second, there are parallel opportunities for human geographers to explore the economic, political, social, and cultural dimensions of the human environment and of human environmental change. But there are even greater opportunities for integrated geographers, who focus on two-way interactions between the biophysical environment and people, including, amongst other topics, the perception and mitigation of natural hazards, pollution, and disease; land degradation and restoration; the exploitation and sustainability of resources; the conservation and preservation of biodiversity, geo-diversity, and heritage; and the human dimensions of local, regional, and global environmental change.

There are parallel opportunities associated with the core concepts of space and place. Technical advances in describing, monitoring, and analysing spatial variation over the Earth's surface by EO and GIS have matured to the extent that the focus of attention is increasingly able to turn towards the scientific principles, modelling, and theory, rather than the techniques themselves. This is part of the 'modernization of geography', with major implications for both its intellectual development and its applications. At the other end of the spectrum of geographical interests, the many profoundly different ways of interpreting and theorizing about place present geography with multiple new opportunities.

Some have suggested that the trend to globalization is signalling the end of traditional geography. They would say that because a world dominated by global-scale processes might seem to diminish the importance of our traditional interest in local and regional considerations. In physical geography, global processes, such as climate change and the carbon cycle, including the impact of 'greenhouse gas' emissions on global warming, are increasingly setting the research agenda. Similarly, rapid communication, corporate business, and international agencies are leading to a new global human geography. However, a major thrust in contemporary human geography has been to question the power of universal processes and to focus on differences, diversity, and the plurality of ways in which people react to and initiate change. There are opportunities opening up from both 'physical' and 'human' globalization in relation to the interplay between scales ranging from the local to the global. The local impacts of global change and the global impact of local events are intimately related. Whether these relate to global warming, a tsunami, terrorism, or a financial crash, they illustrate the continuing importance and relevance of space, place, and environment in the changing world.

Present geography's weaknesses

With such an array of strengths and opportunities, in pure and applied research and in education, it is tempting to say that geography has never had it so good! However, the situation is not as optimistic as it would appear at first sight. Some of the parallel weaknesses and threats are hinted at by Sally Eden in relation to the environmental theme:

Geography took its eye off the environmental ball for the first half of the twentieth century, and then got caught on the hop. Environmental concern in the 1970s found Geography fragmented, unprepared and perhaps unwilling to take a leadership role. Although geographers in Britain and elsewhere have explored a range of environmental themes since then, today 'the environment' is everywhere but nowhere in geographical research. The resulting work has been lively and varied, but in the end somewhat shapeless.

S. Eden, 'People and the Contemporary
Environment' (2003)

She is referring to the internal divisions that have developed in geography while other disciplines have become increasingly interested in environmental topics and new integrated 'environmental' sciences of global change, Earth-systems analysis, sustainability science, and the like have been emerging. Similar weaknesses or threats can be identified in relation to geography's other core concerns of space and place: though geographers have also made important contributions by taking the opportunity to work cooperatively with these other disciplines.

The nature and importance of geography are not well understood. The differences between physical and human geography and the shared core of concepts can be confusing, and the bridging role of geography between the sciences and the humanities may be contested. Its breadth has led to the accusation that it is a 'Jack-of-all-trades, master of none'. It has an image problem as well as an identity problem. Geographical research is not always recognized as such. As a critic once remarked, geographers immerse themselves in issues of science and critical theory but, to the general public, geography is all about maps. Understanding of geography outside the discipline often extends no further than the colloquial. Geography also has a much less visible presence in the media than does history or archaeology, for example. Such misunderstandings lead to geography being undervalued not only by the general public but also those in authority within education, academia, industry, and government. Even within the discipline, there is sometimes a lack of communication and understanding between physical and human geographers, or a lack of mutual support regarding the integrity of geography as a whole.

The future of geography

It appears that the divided loyalties of geographers, combined with external perceptions of the discipline, seem to constitute the main threats to geography reaching its full potential. Will geographers build on their strengths and grasp the opportunities, or will they succumb to the weaknesses and threats? We need to organize ourselves better.

The key question for the future is: how should geography focus and organize itself in order to maximize its strengths, make the most of its opportunities, and fulfil its potential? A way to do this is suggested by the simple model of the structure of geography shown in Figure 30. In this model, a set of concentric zones depicts the core and periphery of geography. Integrated geography is shown with densest shading at the centre of the core of geography. Core areas of geography are those where one or more of the core concepts and methods of the discipline form an important component of research or study, whereas peripheral areas are only loosely connected to the core. Beyond the periphery, geography merges with interdisciplinary fields and other disciplines, each of which has its own definable core. All boundaries between zones are shown as broken lines to indicate they are permeable to the flow of ideas, rather than barriers between the different areas of the diagram.

Physical and human geography comprise the two halves of the diagram, whereas the segments can be viewed as particular specializations (such as geomorphology or economic geography, which, for clarity, have not been named). Vertical lines between physical and human geography represent the differences between the two subdisciplines but, significantly, these differences do not extend into the integrated area of the core, which is defined by the interaction of both physical and human geographical elements. Equally significant is the recognition that some parts of each specialism and of the

30. The future of geography as envisaged by our 'integrated development' scenario

subdisciplines of physical and human geography are core; other parts are peripheral; while yet others extend into the interdisciplinary fields.

Possible scenarios for the future development of geography can be proposed in relation to this model of the structure of geography, and three alternative scenarios are considered here:

* the *'laissez-faire'* scenario
* the 'separate-development' scenario
* the 'integrated-development' scenario.

Each scenario places emphasis on different parts of geography's structure – the specialisms, the subdisciplines, or the disciplinary core – hence elements of these three possible futures can be seen in geography today.

According to the *'laissez-faire'* scenario, which in many ways mirrors recent trends, development is uncontrolled and more or less 'anything-goes'. Geography's existing specialisms would thrive under this scenario and numerous new speciality areas would almost certainly continue to emerge. Geographers would also continue to make major contributions to interdisciplinary research, at least in the short term. It could also be argued that accepting such unplanned development is appropriate. After all, this seems to be what is already happening. Why should future possibilities be constrained when the future cannot be predicted? One reason is that further diversification and specialization is likely to lead to further neglect of the core of geography, with more and more geographers working at the periphery of their discipline, or beyond. This distancing of research and teaching activity from the core could mean that key areas of knowledge and understanding – that comprise geography's mission – would thereby be neglected, ultimately leading to the fragmentation of geography and its absorption by other disciplines or new areas of interdisciplinary activity.

A second possible future – the 'separate-development' scenario – envisages that the subdisciplines of physical and human geography will become increasingly autonomous. The two halves of Figure 30 would separate. Since the mid-20th century, the differences between physical and human geography, in terms of subject matter, literature, methods, and philosophical bases, have become more prominent. This scenario merely recognizes, consolidates, and emphasizes these differences. Many of the disadvantages of this scenario are, however, similar to those of the *laissez-faire* scenario. The integrated core of geography, in particular, is likely to be sidelined. In addition, is physical geography or human geography sufficiently coherent to stand alone? The diversity within each of the subdisciplines and past experience of such separations are certainly sufficient evidence to doubt their viability.

The third and final scenario for consideration is the 'integrated-development' scenario. This envisages a regeneration and expansion of the disciplinary core of geography. There is renewed focus on the core concepts and methods. Geographical theory development in a thriving core informs the subdisciplines and the specialisms which, at the same time, are influenced by ideas from outside the discipline. Disciplinary identity is strengthened and there is a more focused external role for geography in relation to interdisciplinary activity and the neighbouring disciplines. Figure 30 attempts to depict this future diagrammatically. The two-way flow of ideas between core and periphery, and the examples of both core and peripheral concerns of geographers, are indicative important features of this scenario. The core is already recognizable and well defined but is not set in stone. Over time it could change to accommodate new dimensions as the discipline continues to modernize.

In our view, the 'integrated-development' scenario provides the best option for geography. If geography is focused on achieving this scenario, it will become more effective and reach its full potential. In a way, it enables us to have our cake and eat it: a sustainable future for the discipline is ensured in which the diverse aspects of geography are interlinked, interdependent, and mutually supportive. Uniquely geographical core concerns are connected to dynamic specialisms that, in turn, contribute to interdisciplinary activity. By ensuring two-way flows between the core, the periphery, interdisciplinary fields, and other disciplines, all parts of the discipline of geography are promoted, while it contributes fully to the continuum of knowledge and its important, unique contribution as a clearly identified single discipline is confirmed. In short, geography's destiny is fulfilled in a multidisciplinary and interdisciplinary world.

References

Here we give the main sources of our examples, quotations, and figures.

E. A. Ackerman, 'Where is a Research Frontier?', *Annals of the Association of American Geographers*, 53 (1963): 435.

D. E. Alexander, 'Natural Hazards on an Unquiet Earth', in *Unifying Geography: Common Heritage, Shared Future*, ed. J. A. Matthews and D. T. Herbert (Routledge, 2004), pp. 206–02.

American Geographical Society and others, *Geography for Life* (National Geographic Research and Exploration, Washington DC, 1994), p. 18.

Arctic Climate Impact Assessment, *Impacts of a Warming Arctic* (Cambridge University Press, 2004).

L. K. Barlow, J. P. Sadler, A. E. J. Ogilvie, and others, 'Interdisciplinary Investigations into the End of the Norse Western Settlement in Greenland', *The Holocene*, 7 (1997): 489–99.

S. Bieri and N. Gerodetti, 'Falling Women – Saving Angels: Spaces in Contested Mobility and the Production of Gender and Sexualities within Early Twentieth Century Train Stations', *Social and Cultural Geography*, 8 (2007): 217–34.

C. Butler, *Postmodernism: A Very Short Introduction* (Oxford University Press, 2002).

J. Caesar, *The Gallic Wars and Other Writings* (Heron Books edition, 1957), p. 1.

N. Castree, 'Economy and Culture Are Dead! Long Live Economy and Culture', *Progress in Human Geography*, 28 (2004): 204–26.

W. Christaller, *Central Places in Southern Germany*, tr. C. W. Baskin (Prentice Hall, 1966).

D. B. Clarke, 'The City of the Future Revisited or, the Lost World of Patrick Keiler', *Transactions, Institute of British Geographers*, 32 (2007): 29–45.

S. Cloete, *A Victorian Son: An Autobiography 1897–1922* (Heron Books edition, [1923] 1972), p. 1.

J. Comaroff, 'Ghostly Topographies, Landscape and Biopower in Modern Singapore', *Cultural Geographies*, 14 (2007): 56–73.

P. J. Crutzen and E. Stoermer, 'The "Anthropocene"', *International Geosphere Biosphere Programme Global Change Newsletter*, 41 (2001): 12–13.

C. Darwin, *The Voyage of the Beagle* (Heron Books edition, [1845] 1968), p. 1.

W. K. D. Davies, 'Globalization: A Spatial Perspective', in *Unifying Geography: Common Heritage, Shared Future*, ed. J. A. Matthews and D. T. Herbert (Routledge, 2004), pp. 189–214.

M. J. Dear and S. Flusty (eds), *The Spaces of Postmodernity* (Blackwell, 2002), p. 2.

J. Diamond, *Collapse: How Societies Choose to Fail or Succeed* (Viking Press, 2005).

S. Eden, 'People and the Contemporary Environment', in *A Century of British Geography*, ed. R. Johnston and M. Williams (Oxford University Press, 2003), pp. 213–43.

T. S. Eliot, *Little Gidding, The Four Quartets, No.4, Part 5* (Tristan Fecit, [1942] 2000), p. 6.

A. S. Fotheringham, C. Brunsdon, and M. Charlton, *Quantitative Geography: Perspectives on Spatial Data Analysis* (Sage, 2000), p. xi.

G. L. Gaile and C. J. Willmott (eds), *Geography in America at the Dawn of the 21st Century* (Oxford University Press, 2003), p. 1.

P. J. Gersmehl, 'An Alternative Biogeography', *Annals of the Association of American Geographers*, 66 (1976): 223–241.

M. H. Glantz, *Currents of Change: Impacts of El Niño and La Niña on Climate and Society* (Cambridge University Press, 2001), p. 138.

D. Gregory, 'Geographies, Publics and Politics', *Progress in Human Geography*, 29 (2005): 182–93.

J. K. Guelke, 'Mrs Gardner's World: Scale in Mormon Women's Autobiographical Writing', *Area*, 39 (2007): 268–77.

T. Hägerstrand, *Innovation Diffusion as a Spatial Process* (University of Chicago Press, 1968).

D. T. Herbert, *The Geography of Urban Crime* (Longman, 1982).

D. T. Herbert, with N. R. Fyfe and D. J. Evans, *Crime, Policing and Place: Essays in Environmental Criminology* (Routledge, 1992).

D. T. Herbert and J. A. Matthews, 'Geography', in *The Encyclopaedic Dictionary of Environmental Change*, ed. J. A. Matthews and others (Arnold, 2001), p. 255.

International Association for Landscape Ecology, IALE mission statement, *IALE Bulletin*, 16 (1998): 1.

J. Keay, *The Royal Geographical Society History of World Exploration* (Hamlyn, 1991), p. 301.

C. G. Knight, 'Geography's Worlds', in *Geography's Inner Worlds: Pervasive Themes in Contemporary American Geography*, ed. R. F. Abler, M. G. Marcus, and J. M. Olsen (Rutgers University Press, 1992), pp. 9–26.

H. Le Bras, 'World Population and the Environment', in *The Earth from the Air*, ed. Y. Arthus-Bertrand (Thames and Hudson, 2005), pp. 47–52.

J. Liu, T. Dietz, S. R. Carpenter, and others, 'Complexity of Coupled Human and Natural Systems', *Science*, 317 (2007): 1513–16.

D. Livingstone, *Missionary Travels and Researches in South Africa*, from *The Oxford Book of Exploration*, ed. R. Hanbury-Tenison (Oxford University Press, [1857] 1993), pp. 178–9.

P. A. Longley and M. J. Barnsley, 'The Potential of Geographical Information Systems', in *Unifying Geography: Common Heritage, Shared Future*, ed. J. A. Matthews and D. T. Herbert (Routledge, 2004), p. 63.

H. J. Mackinder, 'On the Scope and Methods of Geography', *Proceedings of the Royal Geographical Society*, 9 (1887): 141–60.

G. P. Marsh, *Man and Nature, or Physical Geography as Modified by Human Action*, ed. D. Lowenthal (Belknap Press, [1864], 1965), p. 42.

D. Massey, 'Globalisation: What Does It Mean?', *Geography*, 87 (2004): 293–6.

J. A. Matthews, *The Ecology of Recently Deglaciated Terrain: A Geoecological Approach to Glacier Forelands and Primary Succession* (Cambridge University Press, 1992).

J. A. Matthews and P. Q. Dresser, 'Holocene Glacier Variation Chronology of the Smørstabbtindan Massif, Jotunheimen, Southern Norway, and the Recognition of Century- to Millennial-Scale European Neoglacial Events', *The Holocene*, 18 (2008): 181–201.

Intergovernmental Panel on Climate Change, *Climate Change 2007: The Physical Basis* (Cambridge University Press, 2007).

J. A. Matthews and D. T. Herbert, 'Unity in Geography: Prospects for the Discipline', in *Unifying Geography: Common Heritage, Shared Future*, ed. J. A. Matthews and D. T. Herbert (Routledge, 2004), pp. 369–93.

N. Myers, R. A. Mittermeier, C. G. Mittermeier, and others, 'Biodiversity Hotspots for Conservation Priorities', *Nature*, 403 (2000): 853–8.

National Geographic Society, *Almanac of Geography* (National Geographic Society, Washington DC, 2005), p. 10.

F. Oldfield, *Environmental Change: Key Issues and Alternative Approaches* (Cambridge University Press, 2005).

B. Rolett and D. Diamond, 'Environmental Predictors of Pre-European Deforestation on Pacific Islands', *Nature*, 431 (2004): 443–6.

J. Rose and X. Meng, 'River Activity in Small Catchments over the Last 140 ka, North-East Mallorca, Spain', in *Fluvial Processes and Environmental Change*, ed. A. G. Brown and T. A. Quine (Wiley, 1999), pp. 91–102.

T. Saiko, *Environmental Crises* (Prentice Hall, 2001), Chapter 6, pp. 242–72.

Science, 'Review of Harm de Blij's *The Geography Book*' (1995).

O. Slaymaker and T. Spencer, *Physical Geography and Global Environmental Change* (Longman, 1998), p. 7.

V. Smil, 'How Many Billions To Go?', *Nature*, 401 (1999): 429.

Social and Cultural Geography, A collection of papers on lesbian space: 18(1) (2007).

S. Stevens, 'Fieldwork as Commitment', *The Geographical Review*, 91 (2001): 66.

G. Valentine, *Social Geographies: Space and Society* (Prentice Hall, 2001), p. 1.

L. R. Walker and M. R. Willig, 'An Introduction to Terrestrial Disturbances', in *Ecosystems of Disturbed Ground*, ed. L. R. Walker (Elsevier, 1999), pp. 1–16.

S. Whatmore, *Hybrid Geographies: Natures, Cultures and Spaces* (Sage, 2002), p. 116.

G. F. White, 'Geography', in *Encyclopaedia of Global Environmental Change, Volume 3*, ed. I. Douglas (Wiley, 2002), p. 337.

M. Williams, 'The Creation of Humanised Landscapes', in *A Century of British Geography*, ed. R. Johnston and M. Williams (Oxford University Press, 2003), pp. 167–212.

S. W. Woodridge, *The Spirit and Significance of Fieldwork* (Council for Promotion of Field Studies, 1948), p. 2.

Websites

http://www.sasi.groupshef.ac.uk/worldmapper/display.php?selected=169

http://www.sasigroup.shef.ac.uk/worldmapper/about.html

http://www.landscape-ecology.org/about/aboutIALE.htm

http://www.earthobservatory.nasa.gov

http://glcf.umiacs.umd.edu

Geography

Further reading

Here we suggest sources of further information on the general themes that we have introduced in this Very Short Introduction to Geography.

Chapter 1, Geography: the world is our stage

For a fairly recent full account of the history of geography, see D. N. Livingstone, *The Geographical Tradition: Episodes in the History of a Contested Enterprise* (Blackwell, 1992). G. J. Martin and P. E. James, *All Possible Worlds: A History of Geographical Ideas* (Wiley, 1993) is a substantial text that addresses the history of geography in different parts of the world. A. Holt-Jensen, *Geography: History and Concepts* (Sage, 1999) is a much shorter book and this latest edition provides a good overview of the discipline. R. J. Johnston, *Geography and Geographers: Anglo-American Geography since 1945* (Arnold, 1997) provides a similar, well-established overview. P. Haggett's *Geography: A Global Synthesis* (Prentice Hall, 2001), though written as a student text with a strong orientation towards spatial analysis, is also a very good general introduction.

The fundamental importance of geography in education and research is emphasized in the US National Research Council's *Rediscovering Geography: New Relevance for Science and Society* (National Academy

Press, 1997). A collection of short chapters focusing on the core concepts of geography can be found in S. L. Holloway, S. P. Rice, and G. Valentine (eds), *Key Concepts in Geography* (Sage, 2003).

H. de Blij, *Why Geography Matters* (Oxford University Press, 2005) discusses from a geographer's viewpoint, in a very readable account, practical challenges facing the world, including climate change, the rise of China, and global terrorism. A useful book for those thinking about studying geography at university is A. Rogers and H. A. Viles (eds), *The Student's Companion to Geography*, 2nd edn (Blackwell, 2003).

Thorough, up-to-date reviews of the current state of geography as a research discipline are provided in R. Johnston and M. Williams (eds), *A Century of British Geography* (Oxford University Press and the British Academy, 2003); G. L. Gaile and C. J. Willmott (eds), *Geography in America at the Dawn of the 21st Century* (Oxford University Press, 2003); and I. Douglas, R. Huggett, and C. Perkins (eds), *Companion Encyclopedia of Geography: From Local to Global*, 2nd edn (Routledge, 2007). The first is organized loosely around space, place, and environment as three of the core concepts of geography; the second emphasizes environmental dynamics, social dynamics, and the dynamics of social-environmental interactions as key themes of geography's many specialisms; and the last presents a quite comprehensive set of physical and human geographical essays framed by various views of the concept of place.

Chapter 2, The physical dimension: our natural environments

There are many modern comprehensive introductions to the facts and concepts of the subdiscipline of physical geography. Good examples include: P. Smithson, K. Addison, and K. Atkinson, *Fundamentals of the Physical Environment*, 3rd edn (Routledge, 2002); Alan Strahler and Arthur Strahler, *Physical Geography: Science and Systems of the Human Environment*, 3rd edn (Wiley, 2005); and R. W. Christopherson, *Geosystems: An Introduction to Physical Geography*,

6th edn (Pearson Prentice Hall, 2006). A detailed account of the development of physical geography, including current trends, can be found in K. J. Gregory, *The Changing Nature of Physical Geography* (Arnold, 2000). The same author has also edited a four-volume reader of 65 key contributions to the subdiscipline in *Physical Geography* (Sage, 2005), reprinted from their original sources. Regular reviews and updates of progress in the subdiscipline and its specialisms can be found in the journal *Progress in Physical Geography*, published by Sage.

Exemplars of particular approaches, or paradigms, in physical geography include L. B. Leopold, M. G. Wolman, and J. P. Miller, *Fluvial Processes in Geomorphology* (Freeman, 1964); R. J. Chorley and B. A. Kennedy, *Physical Geography: A Systems Approach* (Prentice Hall, 1971); and O. Slaymaker and T. Spencer, *Physical Geography and Global Environmental Change* (Longman, 1998). The recent emphasis on humans as physical geographical agents is well illustrated by A. Goudie in *The Human Impact on the Natural Environment*, 6th edn (Blackwell, 2006); and also by R. Huggett, S. Lindley, H. Gavin, and K. Richardson in *Physical Geography: A Human Perspective* (Arnold, 2004).

The two-volume *Encyclopedia of Geomorphology*, edited by A. Goudie (Routledge, 2004) provides an excellent example of the breadth and depth of research and study within one of the main specialisms of physical geography. Interdisciplinary contributions by physical geographers are exemplified by: J. J. Lowe and M. J. C. Walker, *Reconstructing Quaternary Environments*, 2nd edn (Longman, 1997); F. Oldfield, *Environmental Change: Key Issues and Alternative Approaches* (Cambridge University Press, 2005); W. M. Marsh and J. Grossa, Jr, *Environmental Geography: Science, Land Use and Earth Systems*, 3rd edn (Wiley, 2005); and J. Wiens and M. Moss, *Issues and Perspectives in Landscape Ecology* (Cambridge University Press, 2005). On the methodological side, R. Haines-Young and J. Petch, *Physical Geography: Its Nature and Methods* (Harper and Row, 1986) provides a good, though somewhat dated, introduction to physical geography as a natural environmental science. Some recent alternative philosophical perspectives are introduced in R. Inkpen, *Science, Philosophy and Physical Geography* (Routledge, 2005), and in S. Trudgill and A. Roy (eds), *Contemporary Meanings in Physical Geography* (Arnold, 2003).

Chapter 3, The human dimension: people in their places

G. Benko and U. Strohmayer have edited a helpful collection of articles under the title *Human Geography: A History for the 21st Century* (Arnold, 2004); while Gill Valentine, *Social Geographies: Space and Society* (Prentice Hall, 2001) reviews many of the new approaches to human geography. An older collection is *Human Geography: An Essential Anthology*, edited by J. Agnew, D. N. Livingstone, and A. Rogers (Blackwell, 1996). Other good introductions to the various paradigms of human geography include: P. Cloke, C. Philo, and D. Sadler, *Approaching Human Geography* (Paul Chapman, 1991); R. Peet, *Modern Geographical Thought* (Blackwell, 1998); P. Daniels, M. Bradshaw, D. Shaw, and J. Sidaway (eds), *Human Geography: Issues for the 21st Century* (Pearson, 2005); and P. Cloke, P. Crang, and M. Goodwin (eds), *Introducing Human Geographies*, 2nd edn (Arnold, 2005).

A selection of works on various specialisms within human geography include: E. Sheppard and T. Barnes (eds), *The Companion to Economic Geography* (Blackwell, 2003); R. Potter, T. Binns, J. Elliott, and D. Smith, *Geographies of Development* (Pearson, 2003); M. Woods, *Rural Geographies* (Sage, 2005); and A. Southall, *The City in Time and Space* (Cambridge University Press, 1998). D. Mitchell's *Cultural Geography: A Critical Introduction* (Blackwell, 2000) offers a particular set of insights into the development and current priorities of cultural geography. The collection edited by M. Dear and S. Flusty, *The Spaces of Post-Modernity: Readings in Human Geography* (Oxford, 2002), covers rather more material than the title implies and is useful both for its reflective chapters and for further examples of modern practice.

The journal *Progress in Human Geography*, published by Sage, London, has regular updates on recent work in the various subdivisions of human geography.

D. Harvey's list of contributions to human geography is impressively long and a selection of the major works of this influential human geographer is worthy of close inspection. His early *Explanation in Geography* (Blackwell, 1969) reflected his involvement in the 'quantitative revolution', but his major works have been written from the viewpoint of Marxist structural theory. They include *Social Justice and the City* (Arnold, 1973) and, most recently, new editions of *The Limits to Capital* (Verso, 2006) and *Space and Global Capitalism: Towards a Theory of Uneven Geographical Development* (Verso, 2006).

Chapter 4, Geography as a whole: the common ground

There are surprisingly few books that cover geography as a whole, and even fewer that emphasize truly integrated geography. In our own volume, J. A. Matthews and D. T. Herbert (eds), *Unifying Geography: Common Heritage, Shared Future* (Routledge, 2004), 29 geographers write specifically on the many integrating themes that permeate the discipline of geography. An earlier important work that also explicitly considers the themes that physical geography and human geography have in common is R. F. Abler, M. G. Marcus, and J. M. Olsson (eds), *Geography's Inner Worlds: Pervasive Themes in Contemporary American Geography* (Rutgers University Press, 1992). Another recent source that seriously addresses the scope, coherence, and core concepts of geography is N. Castree, A. Rogers, and D. Sherman (eds), *Questioning Geography: Fundamental Debates* (Blackwell, 2005).

The following provide further details on the five areas of integrated geography that we discuss in this chapter. Traditional and modern views on regional geography are exemplified, respectively, by R. E. Dickinson, *Regional Concept: The Anglo-American Leaders* (Routledge and Kegan Paul, 1976), and R. J. Johnston, G. Hoekveld, and J. Hauer (eds), *Regional Geography: Current Developments and Future Prospects* (Routledge, 1990). Recent books on historical geography include R. A. Butlin, *Historical Geography: Through the Gates of Space and Time* (Arnold, 1993); B. Graham and C. Nash, *Modern Historical Geographies* (Prentice Hall, 2000); and A. R. H. Baker, *Geography and History:Bridging the Divide* (Cambridge University Press, 2003).

Interaction between people and their environment was used by I. Douglas, R. Huggett, and M. Robinson (eds) as the overarching theme of the first edition of their *Companion Encyclopedia of Geography: The Environment and Humankind* (Routledge, 1996), while the ever more pressing theme of global environmental change has been taken up in a large number of books, including B. L. Turner II, W. C. Clark, R. W. Kates, and others (eds), *The Earth as Transformed by Human Action: Global and Regional Changes in the Biosphere over the Past 300 Years* (Cambridge University Press, 1990); J. R. Mather and G. V. Sdasyuk (eds), *Global Change: Geographical Approaches* (University of Arizona Press, 1991); and A. Goudie (ed.), *Encyclopedia of Global Change* (Oxford University Press, 2002). The scope of landscape geography can be gauged from: P. Adams, I. Simmons, and B. Roberts, *People, Land and Time: An Historical Introduction to the Relations Between Landscape, Culture and Environment* (Arnold, 1998); I. White, *Landscape and History* (Reaktion, 2002); and L. Head, *Cultural Landscapes and Environmental Change* (Arnold, 2000).

Chapter 5, How geographers work

The full range of methods used by geographers is covered in N. Clifford and G. Valentine (eds), *Key Methods in Geography* (Sage, 2003). A Special Issue of *The Geographical Review* 91 (2001) contains 56, mostly idiosyncratic, essays on *Doing Fieldwork*. A modern view of maps and mapmaking that emphasizes the diverse ways in which maps shape human activities is given in J. Pickles, *A History of Spaces: Cartographic Reason, Mapping and the Geo-Coded World* (Routledge,

2004). An illustrated history of maps and cartography is provided by A. Ehrenberg (ed.), *Mapping the World* (National Geographic, 2006), while the modern techniques involved are elaborated in T. Slocum, R. B. McMaster, F. C. Kessler, and H. H. Howard, *Thematic Cartography and Geographic Visualisation* (Pearson Prentice Hall, 2005). The range of numerical techniques employed by geographers is well covered in N. Wrigley and R. J. Bennett (eds), *Quantitative Geography: A British View* (Routledge and Kegan Paul, 1981). M. J. Barnsley, *Environmental Modeling: A Practical Introduction* (CRC Press, 2007) provides an excellent introduction to both the conceptual basis and the practicalities of computer modelling of environmental systems. Earth observation is introduced well by Paul Curran, *Principles of Remote Sensing* (Longman, 1986). The definitive review of GIS is P. A. Longley, M. F. Goodchild, D. J. Maguire, and D. W. Rhind (eds), *Geographical Information Systems and Science* (Wiley, 2001), while the implications as well as the technicalities are ably introduced in N. Schuurman, *GIS: A Short Introduction* (Blackwell, 2004).

A thorough account of the breadth and depth of applied geography is provided by M. Pacione (ed.), *Applied Geography: Principles and Practice* (Routledge, 1999). Good introductions to natural hazards and disasters are I. Burton, R. W. Kates, and G. F. White, *The Environment as Hazard* (Guilford Press, 1993) and two books by D. E. Alexander, *Natural Disasters* (UCL Press, 1993) and *Confronting Catastrophe: New Perspectives on Natural Disasters* (Terra, 2000). A collection of essays published in *Progress in Human Geography*, 29 (2005): 165–93, considers the role of geography in public debate (those by W. Turner and D. Gregory are particularly interesting).

Chapter 6, Geography's present and future

For the first part of this chapter, which considers four examples of the research frontier in geography, appropriate further reading has been listed under 'References'. In relation to the future of geography, the second part of the chapter introduces ideas that have been elaborated in our earlier book: J. A. Matthews and D. T. Herbert (eds), *Unifying Geography: Common Heritage, Shared Future* (Routledge, 2004; see especially the final chapter, pp. 369–93). Alternative views on the future of the discipline of geography are presented in two books edited by R. J. Johnston: *The Future of Geography* (Methuen, 1985) and *The Challenge for Geography: A Changing World, a Changing Discipline* (Blackwell, 1993).

Index

Index

Index

CAPITALISM
A Very Short Introduction
James Fulcher

The word 'capitalism' is one that is heard and used frequently, but what is capitalism really all about, and what does it mean? Fulcher addresses important present day issues, such as New Labour's relationship with capitalism, the significance of global capitalism, and distinctive national models of capitalism. He also explores whether capital has escaped the nation-state by going global, emphasizing that globalizing processes are not new. He discusses the crisis tendencies of capitalism, such as the Southeast Asian banking crisis, the collapse of the Russian economy, and the 1997–1998 global financial crisis, and asks whether capitalism is doomed. The book ends by asking whether there is an alternative to capitalism, discussing socialism, communal and cooperative experiments, and the alternatives proposed by environmentalists.

http://www.oup.co.uk/isbn/0-19-280218-6

MARX
A Very Short Introduction
Peter Singer

Peter Singer has succeeded in identifying the central vision that unifies Marx's thought. He thus makes it possible, in remarkably few pages, for us to grasp Marx's views as a whole, rather than as an economist or a social scientist. He explains alienation, historical materialism, the economic theory of Capital and Marx's ideas of communism in plain English, and concludes with an assessment of Marx's legacy.

> 'An admirably balanced portrait of the man and his achievement.'
>
> **Philip Toynbee,** *Observer*

www.oup.co.uk/isbn/0-19-285405-4

GLOBAL WARMING
A Very Short Introduction
Mark Maslin

Global Warming is one of the most controversial scientific issues of the twenty-first century. This is a problem that has serious economic, sociological, geopolitical, political, and personal implications.

This *Very Short Introduction* is an informative, up-to-date, and readable book about the predicted impacts of global warming and the surprises that could be in store for us in the near future. It unpacks the controversies that surround global warming, drawing on material from the recent report of the Intergovernmental Panel on Climate Change (IPCC), and for the first time presents the findings of the Panel for a general readership. The book also discusses what we can do now to adapt to climate change and mitigate its worst effects.

http://www.oup.co.uk/isbn/0-19-284097-5

D0001623